F9F Panther Cougar

in action

by Jim Sullivan
illustrated by Don Greer

squadron/signal publications

(Cover) The first U.S. Navy jet-to-jet air kill came on 9 November 1950 when LCdr. W.T. Amen, Skipper of VF-111 shot down a North Korean MiG-15. Launching earlier from the USS Philippine Sea, Amen's division of F9F-2B's were out on escort work when they were attacked by MiG-15's who flew right into their formation. With an assist by his wingman Ens. George Holloman, LCdr. Amen downed the Soviet-built swept wing jet within sight of the Yalu River.

F9F-2 (123526) in markings of VMF-311. USMC Air Museum at MCAS Quantico, VA. May 1978. (Jim Sullivan)

COPYRIGHT © 1982 SQUADRON/SIGNAL PUBLICATIONS, INC.
1115 CROWLEY DRIVE, CARROLLTON, TEXAS 75011-5010

All rights reserved. No part of this publication may be reproduced, stored in a retrieval system or transmitted in any form by any means electrical, mechanical or otherwise, without written permission of the publisher.

ISBN 0-89747-127-X

If you have any photographs of the aircraft, armor, soldiers or ships of any nation, particularly wartime snapshots, why not share them with us and help make Squadron/Signal's books all the more interesting and complete in the future. Any photograph sent to us will be copied and the original returned. The donor will be fully credited for any photos used. Please indicate if you wish us not to return the photos. Please send them to: Squadron/Signal Publications, Inc., 1115 Crowley Dr., Carrollton, TX 75011-5010.

Dedication

To all the fighting men of this Nation who cared enough to give their very best.

Acknowledgements

The author wishes to thank the following people for their assistance in providing photos and details that helped put this project together: Hal Andrews, Roger Besecker, Peter Bowers, Robert Carlisle, Tom Curry, Lou Drendel, Bob Esposito, Bill Frierson, M/Sgt. W.F. Gemeinhardt-USMC, Cdr. H.R. Gildcrest-USNR, Ronald Gerdes, Richard Hill, Clay Jansson, Bill Larkins, Bob Lawson, Don Linn, Peter Mancus, Paul McDaniel, Walt Ohlrich, Ron Picciani, Schony Schonenberg, Art Schoeni, Bob Stuckey, William Swisher, Lawrence Webster, Grumman Aerospace Corp., the National Archives and the USMC Air Museum at MCAS Quantico, VA. A special note of appreciation is in order for the help of my son, Jim Jr., who assisted me with the layout preparation of this publication.

F9F-2B (123633) of VF-191 *Octane Sniffer* heads back to USS Princeton after a strike over Korea. 5 April 1951. (USN via Tom Curry)

B
F9F-2B
NAVY
123833
NAVY
104

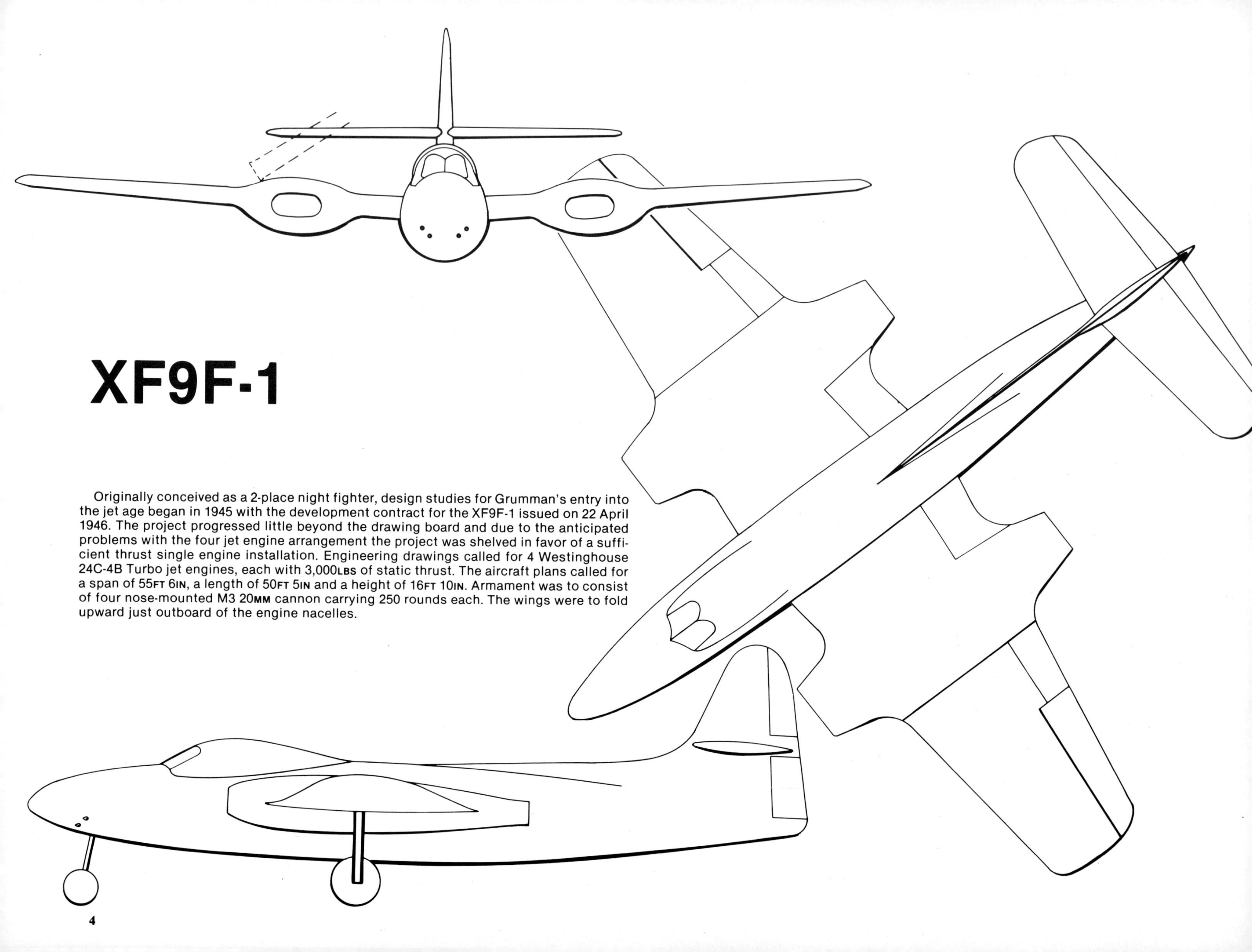

XF9F-1

Originally conceived as a 2-place night fighter, design studies for Grumman's entry into the jet age began in 1945 with the development contract for the XF9F-1 issued on 22 April 1946. The project progressed little beyond the drawing board and due to the anticipated problems with the four jet engine arrangement the project was shelved in favor of a sufficient thrust single engine installation. Engineering drawings called for 4 Westinghouse 24C-4B Turbo jet engines, each with 3,000LBS of static thrust. The aircraft plans called for a span of 55FT 6IN, a length of 50FT 5IN and a height of 16FT 10IN. Armament was to consist of four nose-mounted M3 20MM cannon carrying 250 rounds each. The wings were to fold upward just outboard of the engine nacelles.

F9F Development

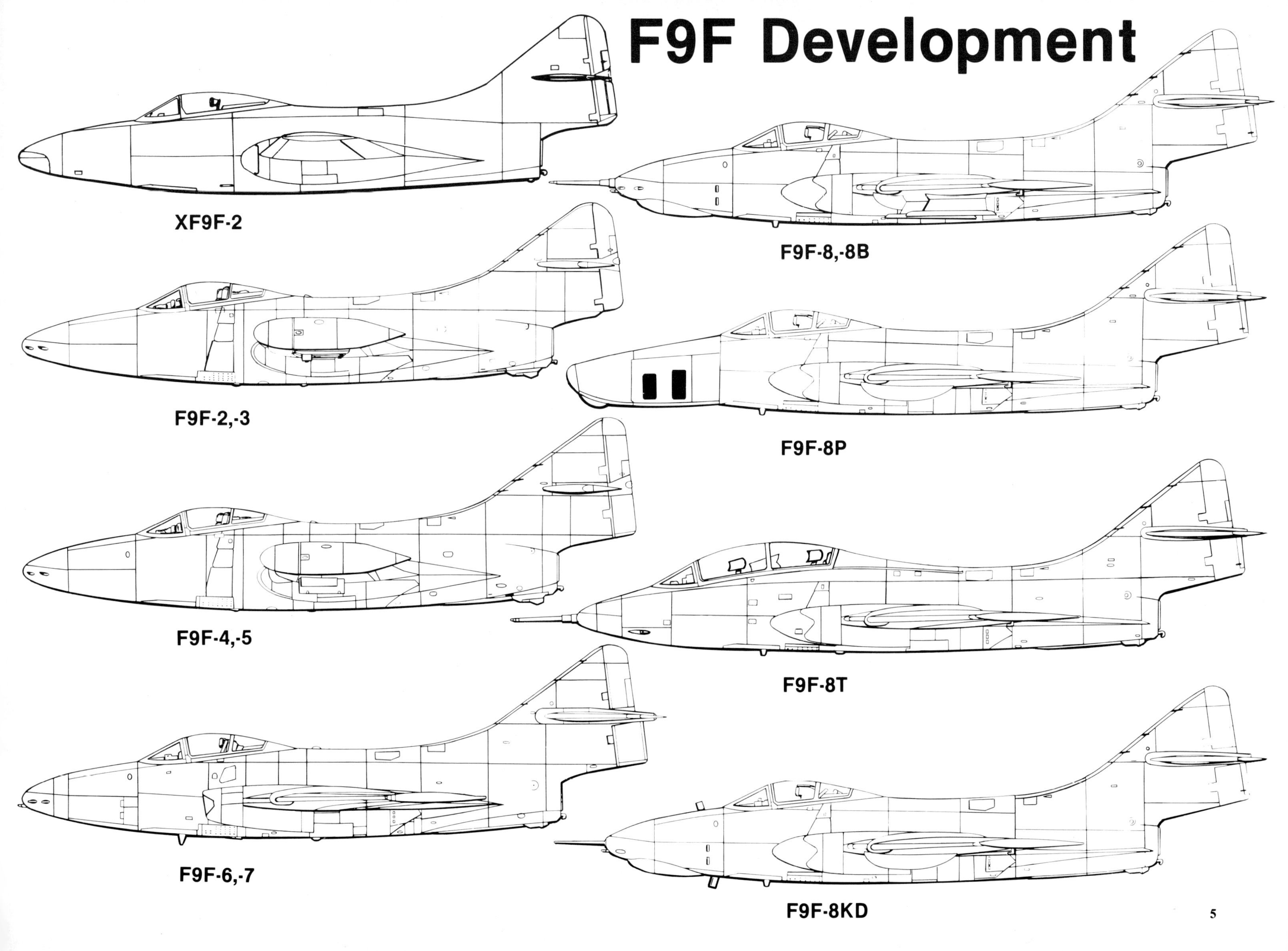

XF9F-2

The XF9F-2 represented Grumman's entry into the then-new field of Jet Aviation. Other U.S. Navy jet prototypes had already flown, i.e., the McDonnell FH-1 Phantom in early 1945, the North American FJ-1 Fury in September 1946, the Vought F6U-1 in October 1946 and the McDonnell F2H Banshee in January 1947. The Grumman-built jet was the aircraft that would be produced in the largest numbers. The Panther as it came to be known was designed around the Rolls-Royce Nene turbojet which developed 5000LBS of thrust. Under a licensing agreement worked out with the British company, the Nene was manufactured in the U.S. by Pratt & Whitney and designated the J42. Grumman's design was engineered for ease of maintenance and introduced a sliding drawer concept for the nose section which just forward of the cockpit extended 3½ feet on rails when released by a cockpit lever and gave easy access to the four 20MM cannon armament section and other nose-mounted systems. On the aft of the XF9F-2, the entire tail section disconnected from the fuselage at the wing flap line. For comparative purposes, the 120 hour check on the jet engine could be completed in half the time it took for similar work on prop-powered aircraft.

A unique feature incorporated on the Panther was a system referred to as **Droop Snoop**. This permitted the forward six inches of the wing leading edge to move downward in conjunction with the wing flaps changing the camber of the wings enabling operation either at high speed with the system retracted or extended for low speed as required for carrier landings. From the formal presentation of the design to the Navy in August of 1946, it took only 15 months for the Panther to take to the sky. Grumman's test pilot C.H. Meyers flew BuNo 122475 for the first time on 24 November 1947. Two airframes (122475 and 122477) carried the XF9F-2 designation.

Some problems did occur with the new design but were dealt with and resolved and the end result was that the U.S. Navy and Grumman reaped the benefits of some eleven years of production before the last in the line of the F9F series was rolled out in 1958.

XF9F-2 (122475) just outside hangar door at Grumman plant, Bethpage, L.I., NY. November 1947. (Grumman via Hal Andrews)

The second XF9F-2 (122477) flown by Grumman test pilot C.H. Meyers sparkles in its Al-clad finish as it banks over Long Island sound. 16 August 1948. (Grumman via Cdr. H.R. Gildcrest, USNR)

XF9F-2 (122475) with Meyers at the controls during taxi tests at the Grumman plant. 20 November 1947. (Grumman)

F9F-2 Cockpit Arrangement

Left Console

1. "G" Suit Valve Control
2. Radio Altimeter Selector Switch - Airplane Serial No. 122572 and subsequent
3. Oxygen Regulator
4. Air Starts Switch "ON-OFF" Buttons
5. Wing Tip Tank Fuel Dump Switch
6. Tip Tank Fuel Flow Warning Lights
7. Right Wing Tip Fuel Tank Switch
8. Left Wing Tip Fuel Tank Switch
9. J33-A-8 Engine Control Switch Panel - F9F-3 Airplanes
10. J33-A-8 Engine Emergency Fuel System Check Switch.
11. J33-A-8 Engine Starting Master Switch - Airplanes Serial No. 122573 and Subsequent
12. J33-A-8 Engine Ground Starts Switch
13. J33-A-8 Engine Starting Fuel System Selector Switch
14. J33-A-8 Engine Starting Switch Guard (Shown In Raised Position)
15. Ejection Seat Emergency Control
16. Wing Flaps Control
17. A.F.C.S. Controls
18. Aileron Booster Emergency "ON-OFF" Control
19. Fuel Master Switch
20. Airplane Fuel System Booster Pump Cutout Switch Button
21. Water Injection Quantity Indicator
22. Wheel and Flap Position Indicator
23. Water Injection Pressure Indicator
24. Water Injection Switch
25. Wing Leading Edge Position Indicator
26. Dive Brake Control
27. Throttle (with Microphone Button)
28. Aileron Trim Tab Control
29. Rudder Trim Tab Control
30. Relief Tube
31. Elevator Tab Control
32. Emergency Brake "T" Handle
33. Personnel Gear Receptacle
34. J42-P-4 or 6 Engine Fuel System Selector Switch
35. J42-P-4 or 6 Engine Emergency Fuel System "ON" Indicator Light
36. J42-P-4 or 6 Engine Starting Switch Guard (Shown In Raised Position)
37. J42-P-4 or 6 Engine Starting Master Switch - Airplanes Serial No. 122572 and Subsequent
38. J42-P-4 or 6 Engine Ground Starts Switch
39. J42-P-4 or 6 Engine High Pressure Cock Switch
40. J42-P-4 or 6 Engine Top Fuel Pump Warning Light
41. J42-P-4 or 6 Engine Bottom Fuel Pump Warning Light
42. J33-A-8 Engine Fuel System Selector Switch
43. J33-A-8 Engine Emergency Fuel System "ON" Indicator Light

Left Console

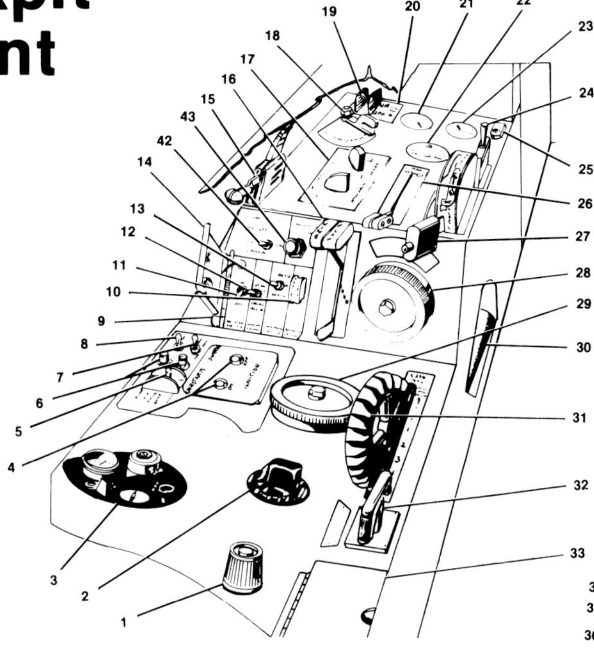

Main Panel

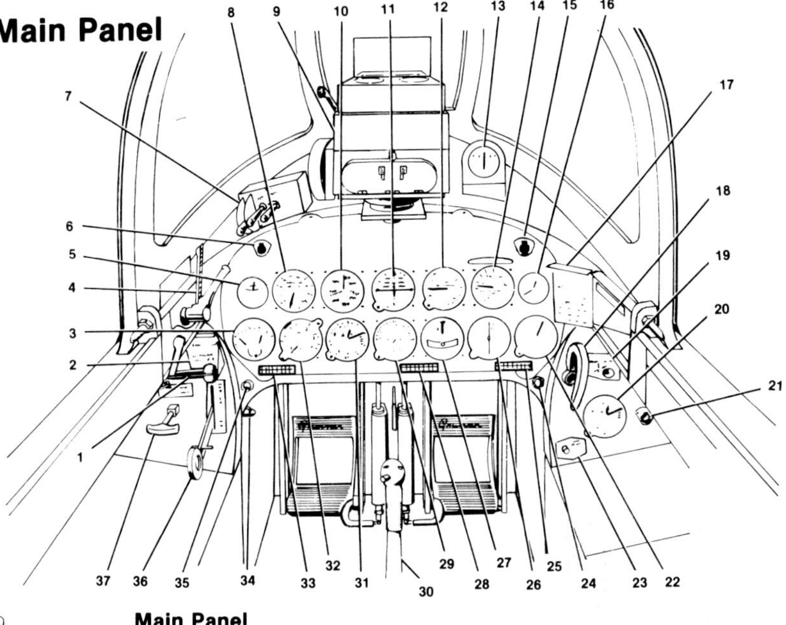

Main Panel

1. Canopy Control (Normal Hydraulic)
2. Take-Off Check-Off List
3. Tail Pipe Temperature Indicator (See Note)
4. Canopy Control (Emergency Air)
5. Fuel and Oil Pressure Indicator
6. Low Fuel Boost Pressure Warning Light
7. Armament Switch Panel
8. Tachometer
9. Mk. 8 Mod. 0 Gunsight (A.F.C.S.)
10. Maximum Allowable Air Speed Indicator
11. Gyro-Horizon
12. Rate-of-Climb Indicator
13. Stand-By Compass
14. Fuel Quantity Indicator
15. Low Level Fuel Warning Light
16. Clock
17. Landing Check-Off List
18. Arresting Hook Control
19. Arresting Hook Indicator Light
20. Cabin Altimeter
21. Sliding Nose "UNLOCKED" Warning Light
22. Accelerometer (See Note)
23. Arresting Hook Raising Switch Button
24. Fire Warning Light
25. Compass Correction Card
26. Radio Compass - Airplane Serial No. 122572 and Subsequent (See Note)
27. Turn and Bank Indicator
28. Compass Correction Card
29. G-2 Compass
30. Control Stick (Trigger Button On Grip)
31. Altimeter
32. Radio Altimeter Indicator - Airplane Serial No. 122572 and Subsequent (See Note)
33. Air Speed Correction Card
34. Landing Gear Down Lock Solenoid Emergency Manual Release
35. Landing Gear "UNLOCKED" Warning Light
36. Landing Gear Control Lever
37. Landing Gear Emergency "T" Handle.

Right Console

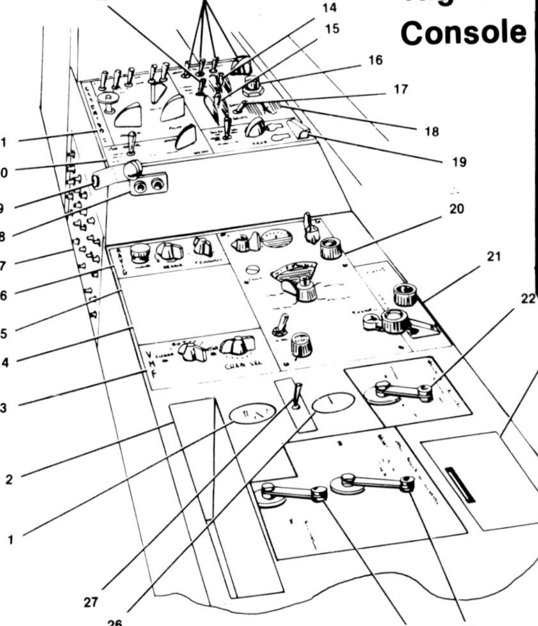

Right Console

1. Hydraulic System Pressure Gage
2. Map Case
3. C-115/ARC-1 Control Panel - Airplanes Serial No. 122572 and Subsequent
4. Location of C-268/ARC-19 Control
5. Location of C-115/ARC-1 Control Panel - Airplanes Serial No. 122560 through 122571 Inclusive
6. C-116/ARR-2A Control Panel - Airplanes Serial No. 122572 and Subsequent
7. Circuit Breaker Reset Button Panel
8. Electrical Test Pin Jacks
9. Seat Height Control
10. Interior Lights Switch Panel
11. Exterior Lights Switch Panel
12. G-2 Compass Switch
13. Cabin Pressurizing System Control Panel - Airplanes Serial No. 122572 and Subsequent
14. Generator "ON-OFF" Switch
15. Battery Switch
16. Generator Warning Light
17. Pitot Heater Switch
18. Volt-Ammeter
19. C-119/APX-1 or -1A Control Panel - Airplanes Serial No. 122572 and Subsequent
20. C-149/ARN-6 Control Panel - Airplanes Serial No. 122572 and Subsequent
21. Radio Master Switch Panel - Airplanes Serial No. 122572 and Subsequent
22. Hydraulic System Shut-Off Control (Combat - Normal)
23. Spare Lamp Container
24. Wing Lock Lock Control
25. Wing Folding Control
26. Hydraulic Auxiliary Pump Pressure Gage
27. Hydraulic Emergency Auxiliary Pump Switch

F9F-2

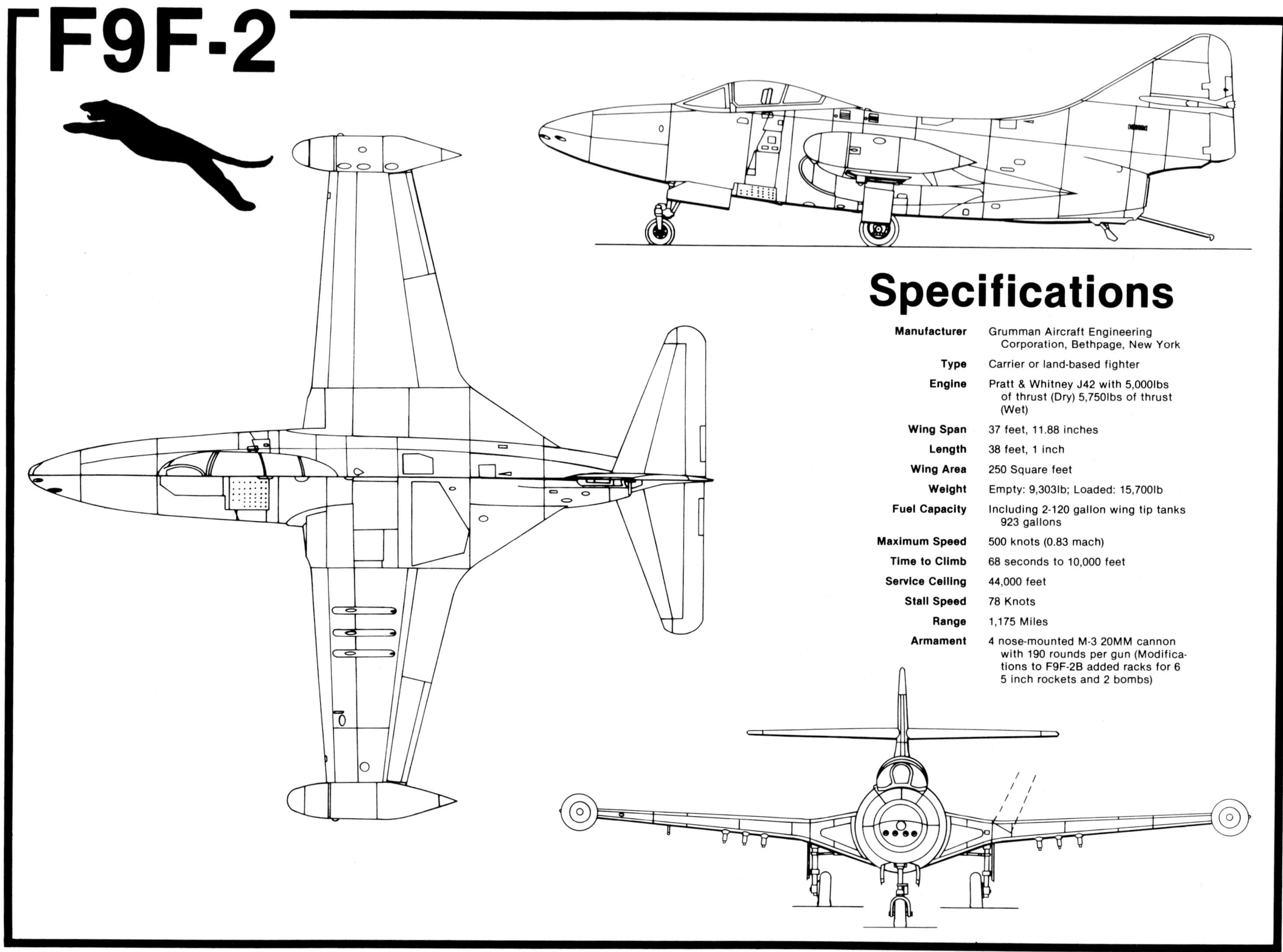

Specifications

Manufacturer	Grumman Aircraft Engineering Corporation, Bethpage, New York
Type	Carrier or land-based fighter
Engine	Pratt & Whitney J42 with 5,000lbs of thrust (Dry) 5,750lbs of thrust (Wet)
Wing Span	37 feet, 11.88 inches
Length	38 feet, 1 inch
Wing Area	250 Square feet
Weight	Empty: 9,303lb; Loaded: 15,700lb
Fuel Capacity	Including 2-120 gallon wing tip tanks 923 gallons
Maximum Speed	500 knots (0.83 mach)
Time to Climb	68 seconds to 10,000 feet
Service Ceiling	44,000 feet
Stall Speed	78 Knots
Range	1,175 Miles
Armament	4 nose-mounted M-3 20MM cannon with 190 rounds per gun (Modifications to F9F-2B added racks for 6 5 inch rockets and 2 bombs)

F9F-2 and F9F-3 Panthers move down the Grumman assembly line mating wings to the fuselages. Note the Glossy Sea Blue wing paint complete with the National insignia. The forward sliding leading edge slats are left unpainted in their natural aluminum finish. Some of the pre-painted sections of the wings, and the canopy area are protected by taped on paper. Bethpage, L.I., NY mid-1949. (Hal Andrews)

F9F-2,-2B

The U.S. Naval Air Test Center, Flight Test Division report on the Panther describes the F9F-2 as a single-seat, single-engine, jet-propelled low wing monoplane designed as a land and carrier-based fighter. The first production Panther flew on 24 November 1948 and was powered by a Pratt & Whitney J42-P-6 engine producing 5,750LBS of thrust. The F9F-2 was produced with two 120 gallon non-jettisonable wing tip fuel tanks, a pressurized air conditioned cockpit, an ejection seat, automatic radio direction finder and a quick-release bubble canopy. All things considered, a very advanced design for a 1948 vintage fighter. Armament for the F9F-2 was four 20MM M-3 cannon located in the nose. These guns carried a total of 760 rounds provided in four ammuntion boxes connected to the guns by feed chutes. When the Panther entered combat, the fighter/bomber concept called for the addition of hardpoints on the wing to carry bombs and rockets. These features were added and that modification was designated the F9F-2B. In short order, all F9F-2 aircraft were retrofitted with underwing racks and rocket launch stubs and the F9F-2B designation was dropped and reverted back to the F9F-2.

Because of the short landing gear arrangement that made the Panther squat in a tail-low position, a unique problem was encountered. During a prolonged engine run-up on board a carrier, the caulking between the flight deck planks would be burned...while on an asphalt runway, the asphalt would melt. Quite understandable with a tailpipe exhaust temperature of approximately 1,000 degrees Fahrenheit. These problems were solved by a concrete run-up area for air stations and an exhaust deflector for carriers. The F9F-2 carried 923 gallons of fuel and had a 38FT wingspan, an overall length of 37FT 10IN and a height of 11FT 4IN. Top gross weight was 15,700LBS fully fueled and armed. At 22,000FT, the Panther would fly at 526MPH. Initial combat use showed the early F9F-2s to be underpowered. On calm days with the carrier making full speed, the Panther required a catapult launch and could just get airborne with loads of 20MM ammo or a pair of 100LB bombs or a couple of rockets. The mission would last approximately 90 minutes because of fuel consumption. The F9F-2B carried the Mark-51 bomb rack just outboard the landing gear and could accommodate up to a 500LB bomb on each of the two racks. Each wing had three launch stations for rockets or could mount up to a 250LB bomb on each of the six stations.

Although VF-51 was the first Panther squadron, VF-111 was the first West Coast squadron to receive the F9F-2. (VF-51 received F9F-3s). A former VF-111 Panther pilot, Ronald Gerdes gives the following information about the Grumman-built Panthers assigned to Task Force 77 off the coast of Korea...Targets were assigned to various squadrons and consisted of three primary missions: Interdiction (fighter/bomber missions against ground targets), C.A.P. (combat air patrol - orbiting the fleet armed with only 20MM guns) and Escort (accompanying unarmed photo-reconnaissance aircraft to their target). The flight schedule would come out the night before...the C.A.P. aircraft would launch just before dawn, the strike aircraft were generally launched just as the sun started to break the horizon. The Panthers flew in groups of four aircraft (two section leaders and two wingmen) heading for the target area. After the target was identified, bombs were dropped first, then rockets, then the 20MM ammunition was expended...after that, the immediate objective was to get out and back to the ship.

Typical flight operations would find the plane handlers attaching the tow bar to align the Panther on the catapult...the section leader on the port cat, the wingman on the starboard. After the aircraft was tensioned on the cat, final checks were made in the cockpit and the cat officer would come out and wind you up...full power, 100%RPM with engine exhaust gas temperature at 1,650 degrees Fahrenheit...stiff-arm the stick with the right elbow in the gut so that when the cat fired, the stick wouldn't come back and give up-elevator...the left hand would push the throttle up and make a fist...push that into the throttle and stiff-arm so during launch the throttle wouldn't come back, which would mean disaster. Then we'd salute, put our head back into the headrest...the cat officer would signal to his man on the catwalk who would push the button that would light up a

light in the catapult room under the forward part of the flightdeck. The guy down there would throw the lever which allowed the 3,500LB hydraulic pressure to push the piston which through a bunch of cables would hurl the F9F off the bow of the ship...a wild ride. Off we'd go, joining up two and two making a division of four...call into the controller and vector off to the beach while climbing up to 10,000FT (the usual altitude used going to and from the target). Once the route was established, the aircraft would go into trail formation keeping the lead aircraft in sight. Trail formation allowed the pilots to search for targets of opportunity. At the target, several passes were generally required to expend bombs, rockets and 20MM ammo...after the ordnance was dropped, the Panthers would climb back to altitude and head for the carrier...arriving there pilots would receive either a **Dog** or **Charlie** signal...**Dog** meant to orbit, **Charlie** meant clear to land. The tailhook was extended by grabbing the lever in the cockpit and making two or three pulls like a windowshade. Each stroke would send the hook further down the track and when it reached the track end, it would drop into the "down" position. Generally the pilots would release the hooks when they reached the "feet wet" position (back over the ocean) so any problems with the tailhook could be seen then such as flak damage or an overheated rail (because of the hot tailpipe immediately above)...this early hook deployment allowed a divert option to a ground base while at the coast rather than flying out to the carrier to learn too late of a hook problem. With a damaged tailhook and not enough fuel to make a land base, the only options were to ditch or to take the barrier, both were things to avoid if at all possible. A normal landing approach meant coming down in *idle*, approaching on the starboard side of the carrier which would be heading into the wind. Careful attention was always necessary to avoid flying into the landing pattern of accompanying carriers. When the leader of the echelon passed the bow of the ship, he would break left at about 300FT and turn downwind. He was followed at ten second intervals by the other three aircraft in the section. Landing aircraft would lower flaps and gear while strung out on the downwind leg. Approach speed was 114KNOTS down to about 130FT off the water (flightdeck was about 65FT off the water). The canopy was opened to assist a hasty exit in the event of an accident...hold 114KNOTS and turn into the carrier looking for a roger and cut signal...receiving that, the pilot would go to idle on the throttle, drop nose over, flare out over the deck and hopefully pick up the #2 wire and stop before hitting the barrier. A hook on #1 wire while acceptable, meant that the approach was too low with a danger of breaking the hook on the round-down (ramp on the rear deck). After the landing, the hook and the landing wire were quickly disengaged, wings were folded and the Panther taxied forward to clear the deck for other aircraft to land.

Pilots would unstrap as quickly as possible to get out of the aircraft due to the possibility of subsequent landing aircraft coming through the barrier and slamming into the parked planes.

F9F-2 and -2B Panthers were flown by both the Navy and the Marines. From June of 1950 until July of 1953, Panthers flew more than 78,000 combat missions. The record-holding F9F-2 carried and dropped more than 400,000LBS of ordnance, fired over 100,000 rounds of 20MM ammo and wore out 16 guns.

A total of 567 F9F-2/2B Panthers were produced.

F9F-2 (123564) seen here in experimental two-tone finish during tests in 1950. Panther appears to be in standard Navy Glossy Sea Blue with Light Grey undersides. This F9F-2 went on to serve with eight different squadrons and was stricken from the records on 1 February 1957. (Grumman)

F9F-2 (122567) in standard Navy finish flys over Long Island Sound during factory test flight. Note perforated dive flaps located on fuselage bottom under the pilot. 1950. (via Bill Larkins)

U.S. Navy Blue Angels flight demonstration team Panther No. 5 sparkles in the California sunshine. The trim color on all markings is Yellow. c. 1953. (Via Peter Bowers)

VF-123

At the moment of arrest this F9F-2 has the wire pulled tight as it is recovered during a post-Korean cruise aboard the Phillipine Sea. VF-123 which was a former reserve squadron (VF-871 from NAS Oakland) was redesignated on 18 February 1953. During this cruise, fighting 123 received a rating of Excellent. 4 July 1955. (National Archives)

F9F-2 Armament System

1. Aft Cockpit Armor Plate
2. Control Stick Trigger Grip
3. Ranging Control (On Throttle Grip)
4. Bullet Resistant Glass Windshield
5. CG-4 (GGS Recorder)
6. Mk 8 Mod. O Sight Unit
7. Armament Switch Panel (Gun Master, Gun Charging)
8. A.F.C.S. Electrical Control Box (On L.H. Console)
9. Forward Cockpit Armor Plate
10. Gun No. 2 Ammunition Box
11. Gun No. 3 Ammunition Box
12. Gun No. 1 20mm M3
13. Gun No. 2 20mm M3
14. Gun No. 3 20mm M3
15. Gun Blast Shields
16. Gun No. 4 20mm M3
17. Gun No. 4 Ammunition Box
18. Gun No. 1 Ammunition Box
19. Gun Camera
20. Forward Boresighting Rod
21. Rear Boresighting Rod

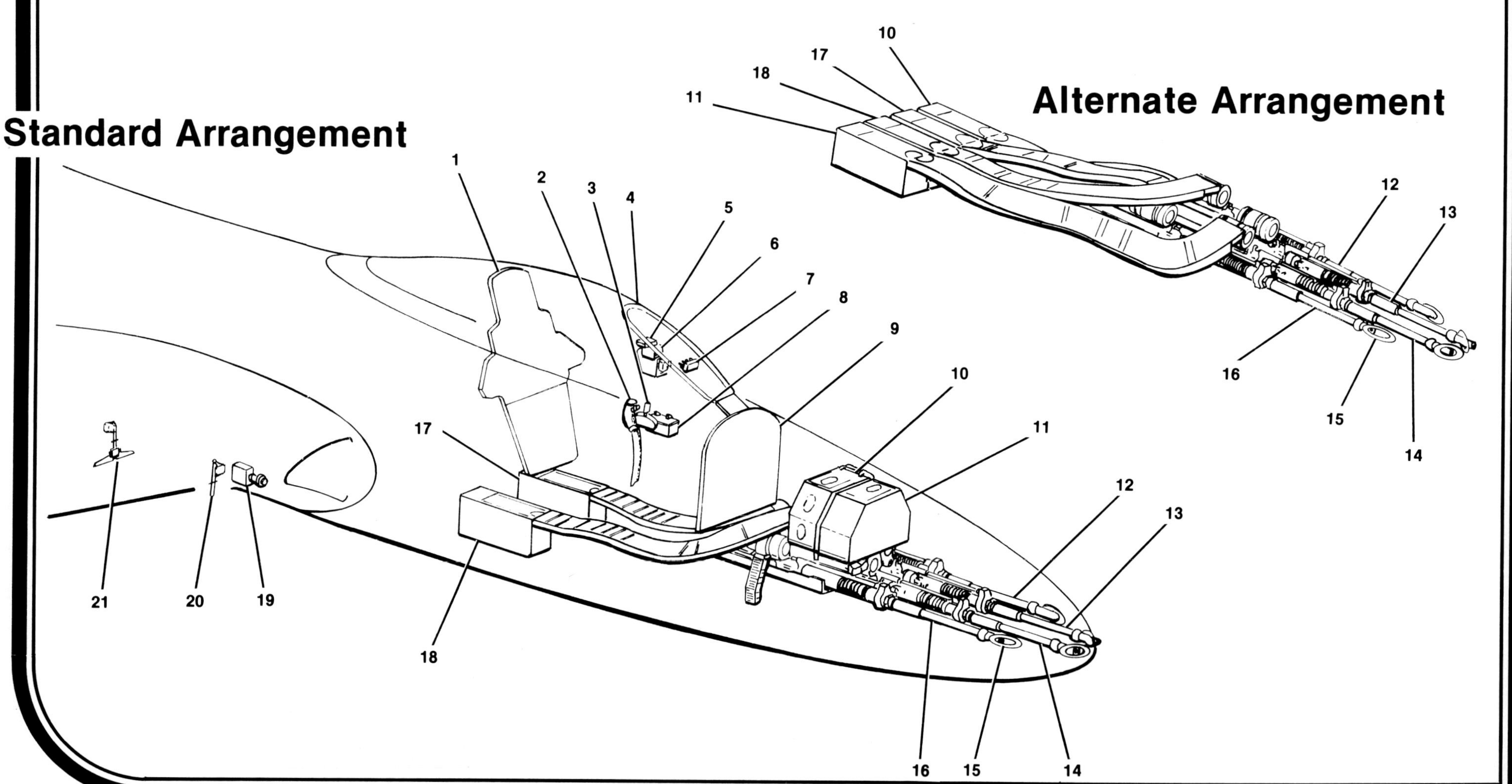

F9F-2 (123497) of VF-93 is hoisted to flight deck for carrier qualifications to be held at sea off the San Diego coast. This Panther has Medium Blue trim on the nose and rudder tip. 1 September 1953. (National Archives)

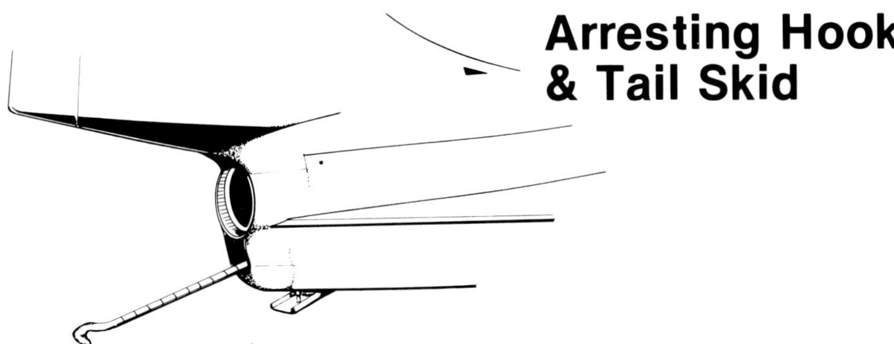

(Above Right) F9F-2 (127087) of VF-61 lifts off from NAS Oceana, VA runway. Aircraft nose and tail trim is White. The pilot is busy cleaning up the bird by retracting gear and bringing up the flaps. 18 May 1952. (National Archives)

(Right) F9F-2 (127210) lets down over Task Force 77 lining up for recovery aboard the USS Boxer off Korea. Trim paint is Yellow. 4 July 1952. (National Archives)

F9F-2 of VF-111 just moments away from landing aboard the USS Philippine Sea. Aircraft is trimmed in Red. Note the six inch downward deflection of the "droop-snoop" wing leading edge slats. November 1950. (Ronald Gerdes)

F9F-2B (123656) of VF-781 seen over Oahu, Hawaii only days before deploying aboard the USS Bon Homme Richard for Korean combat. VF-781 from NAS Los Alamitos was recorded as the first squadron in the U.S. to volunteer for active duty after the outbreak of the Korean conflict. May 1951. (National Archives)

Dive Brake

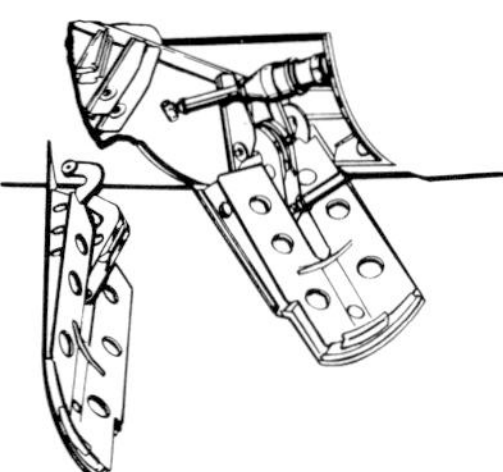

Business end of four 20mm cannon positioned in the nose of a F9F-2B at MCAS Quantico, VA. May 1978. (Jim Sullivan)

Winter snow storm begins to cover flight deck and aircraft aboard the USS Philippine Sea off Korea. 200 series numbered Panthers are VF-112, 100 series are VF-111. 15 November 1950. (National Archives)

F9F-2 (123600) of VF-191 from the USS Princeton heads inland over the rugged mountains of Korea. VF-191 was credited with being the first to use the Panther armed with bombs — on that mission, a railroad bridge was demolished with eight direct hits. 29 May 1951. (National Archives)

Lt. Theodore Nielson of VF-721, a reserve outfit from NAS Glenview, flies F9F-2B (123649) on a photo-escort mission over Korea. Trim on this Panther is Red. 19 June 1951. (National Archives)

Ens. Ronald Gerdes of VF-111 pauses for a photo before climbing into his F9F-2 aboard the USS Philippine Sea. During the winter weather, the pilots would wear *poopy suits*...long-john underwear covered by a very uncomfortable rubber flight suit. This protected the *poopy suit* from getting snagged or torn. A tear in the suit hastened the entry of freezing water in the case of a crash or ditching. On top of the summer flight suit was a Mae West life jacket, shark-repellent (known to the pilots as barracuda-attractor), flares, target-folders, charts, a Smith & Wesson .38 pistol and holster. Outfitted like this, just getting into the Panther was an interesting experience. (Ronald Gerdes)

F9F-2 (127147) with VF-837 encountered a collapsed nosewheel during a landing aboard the USS Antietam off the Korean coast. The Panther flew again after a day's repair. February 1952. (USN via Hal Andrews)

F9F-2 (127184) usually flown by the C.O. of VF-111, Lt. Cdr. W.T. Amen seen during the return to the USS Philippine Sea after a Korean strike. November 1950. (Ronald Gerdes)

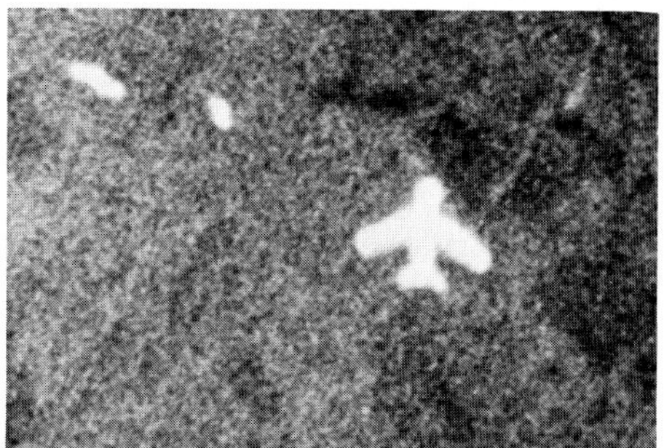

Lt. Cdr. W.T. Amen Scores 1st MiG-15 Jet Kill

A first in the annals of Naval aviation was recorded on 9 November 1950 when Lt. Cdr. Tom Amen C.O. of VF-111 scored a MiG-15 kill with a F9F-2 Panther. This was the first jet-to-jet kill for the U.S. Navy. Launching from the USS Philippine Sea and leading a division of four Panthers, Amen's flight was out to escort ADs and F4Us bombing bridges and powerplants. The Panthers were flying flak suppression, armed with 20MM and light frag bombs to hold the enemy gunners down. All of a sudden they were attacked by MiG-15s, who flew right into the Panther formation, so closely that our pilots were able to identify that the enemy pilots were wearing leather helmets. The MiG-15 that was shot down that day was chased by Cdr. Amen and his wingman George Holloman. The Panthers moved in on the MiG-15, attacking from the rear. It was all the Panthers could do to stay within firing range of the MiG. Amen and Holloman started lobbing 20MM A.P.T. (armor piercing tracer) and H.E.I. (high explosive incendiary) shells. When the H.E.I. would hit the aircraft, they would explode and light up with a bright sparkling appearance. The tracer with this load was for adjusting fire. All of a sudden, the MiG twisted from its straight path, and trailing smoke and flame nosed over into a dive that continued right into the ground. The aircraft impacted within sight of the Yalu River, the border between Korea and China. Both Amen and Holloman returned safely to the ship to a well-done and welcome aboard. An interesting fact is that Amen that day was flying a Panther borrowed from their sister-squadron VF-112. While VF-111 got the credit for the kill, the VF-112 Panther received the MiG-15 kill mark on their F9F.

F9F-2 (127125) of VF-51 returns to the USS Essex after mission over Korea. Earlier in the conflict, VF-51 scored the first two kills (YAK-9s) on 3 July 1950 while flying from the USS Valley Forge. 15 September 1952. (National Archives)

F9F-2 (123554) of VF-151 makes his approach over the USS Boxer. Note the dent in the tip tank. Korea, July 1953. (National Archives)

F9F-2B Ordnance Arrangement

Six 5IN. Standard Air-to-Ground Rockets

Two 1000LB. High Explosive General Purpose Bombs

Six 100LB. G.P. Bombs and Two 500LB. H.E. Bombs

A well-weathered F9F-2 (127210) of VF-151 from the USS Boxer off Korea. Nose trim is Medium Blue. 16 July 1953. (National Archives)

F9F-2 flown by Ens. Chew of VF-34 lands aboard the USS Leyte. Trim on this Panther is Yellow. Note the tailhook just about to engage the #3 wire. 6 August 1951. (National Archives)

Floating high too long, a F9F-2 (123494) of VF-21 lands past the wires, rips through the barrier and is on her way to a fiery crash into other Panthers spotted forward on the deck of the USS Midway. 11 November 1951. (National Archives)

F9F-2 (127207) of VMF-311 at K-3 (Pohang) base, Korea. No. 2 has Red and White trimmed nose...Panthers are being armed with bombs and 20mm. Fall 1952. (USMC)

Being towed down the taxiway prevents this F9F-2B (123602) of VMF-311 from stirring up a dust storm. This armed Panther will be parked in a revetment to await its next mission. Summer 1952 at K-3. (USMC)

Landing Gear Details

Nose Gear

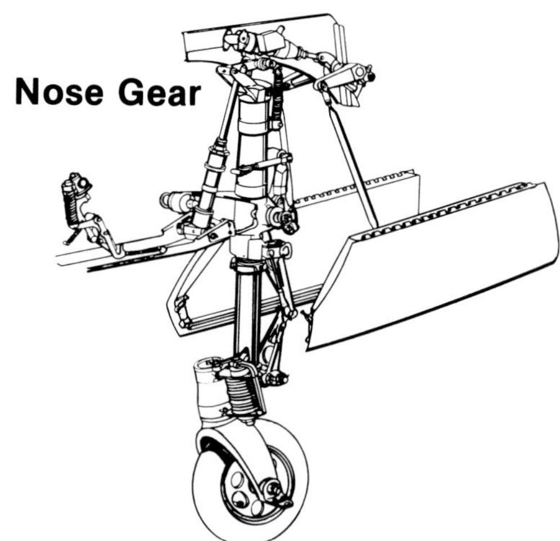

Main Landing Gear

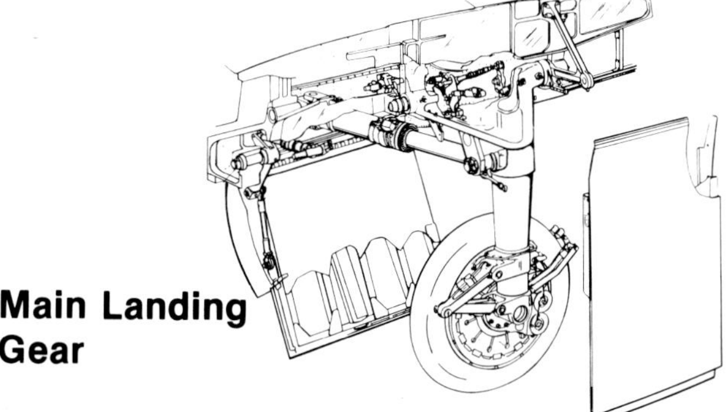

F9F-2B of VMF-311 undergoes thorough maintenance with almost every access panel open. Nose marking is in the form of a Panther head and claw...mostly White with Red outline on eye and back of claw. Korea, 28 June 1951. (USMC)

Lt. John Chalbeck of VF-721 flies his F9F-2B (123713) back to the USS Boxer after a mission over Korea. Nose and rudder trim is Red. 29 August 1951. (National Archives)

(Above Right) F9F-2 of VF-112 marked V-209 makes prolonged attempt to regain airspeed to keep the Panther out of the water...moments after this shot, the aircraft went in off the port side of the USS Philippine Sea off the Korean coast. 3 December 1951. (National Archives)

F9F-2 (123543) of VMF-311 has been rearmed with rockets and 20MM and is being refueled to go back on another strike. Note gun camera location just outboard of air intake. Nose trim is Red and leading edge of wing and tip tank is natural aluminum. Korea, Fall 1952. (USMC)

Wing Fold

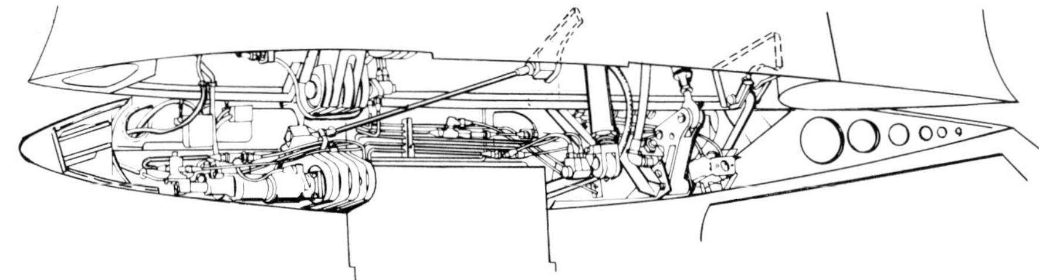

F9F-2 (123507) went off the runway and is hung on an embankment at NAS Knahoe Bay, Hawaii. This Panther was assigned to VMF-232 (Red Devils) c.1953. (Clay Jansson via Peter Mancus)

F9F-2s of VF-111 (127173 & 127191) from the USS Valley Forge landed aboard the USS Philippine Sea. As tradition dictated, visiting aircraft are *decorated* by hands from the host carrier. Panthers seen here are preparing to depart. 31 March 1952. (National Archives)

(Above Left) F9F-2s on flightline of ATU-206. #74 is BuNo 125091 and is trimmed in Red stripes on the tip tanks and solid Red on the nose and rudder and fin tip. Standard Gull Grey top and insignia White bottom. NAS Pensacola, Florida. 30 July 1957. (USMC)

F9F-2KD

A small number of F9F-2P Panthers were converted by the Navy for use as drone director aircraft. The addition of the specialized radio equipment and its number of required antennae completed the modification. Modified Panthers then received a colorful paint job and unit markings.

F9F-2KD (123071) of GMGRU-1 banks over the Pacific off California Coast. c.1959. (Clay Jansson via Peter Mancus)

F9F-2P (123709) of VU-1 over Waikiki Beach, Hawaii. 12 November 1953. (National Archives)

F9F-2P

During the Korean conflict, Navy F9F-2P Panthers, the first Navy photo jets, roared northward through the Korean skys. The F9F-2P was a Navy modification to a basic F9F-2 that substituted an aerial camera and controls for the four 20mm cannon and ammo bay installation. These unarmed but generally fighter escorted Panthers provided aerial photographic intelligence in support of Naval operations. Most of the F9F-2P missions in Korea were flown by pilots of VC-61.

F9F-2P (123706) carries the Red markings of ATU-206 based at Sherman Field at NAS Pensacola. This aircraft logged some combat time in Korea flying with VC-61 off the USS Boxer but is now relegated to advanced training work. 1956. (Peter Bowers)

XF9F-3

The XF9F-3 (122476) was externally identical to the XF9F-2 but was out-performed by them as a result of the difference in powerplants. The XF9F-3 had the Allison J-33-A-8 engine which produced 4,600LBS of thrust. Only one was produced and the idea behind it was for use as a back-up for the possibility that the F9F-2 production would be held up due to engine production delays. Eventually the XF9F-3 was used to carry out carrier acceptance tests.

XF9F-3 (122476) first flew in August of 1948. Seen above in the al-clad finish with the Panther art and wording on the nose over Long Island Sound. 1948. (Grumman)

F9F-3 (122575) of VF-51 lands aboard the USS Boxer during squadron carrier qualifications off the California coast. VF-51 was the first unit to receive Panthers. 13 September 1949. (National Archives)

F9F-3

The Allison-powered F9F-3 featured the same cockpit conveniences as the F9F-2s. Both series of Panthers had hinged doors at the fuselage spine that opened downward by suction to provide additional air under take-off and low-speed flight conditions. Like the -2 Panthers, the -3s had a total fuel capacity of 923 gallons. Inverted flight, like many of the prop jobs, was limited to ten seconds. Due to the better speed of the F9F-2s, most F9F-3s were eventually refitted with Pratt & Whitney J42-P-8 engines and redesignated as F9F-2s...the assigned aircraft BuNo was left unchanged. A total of 54 F9F-3s were produced.

F9F-3 (123031) of VF-24 during strike on enemy coastal targets in Korea. Note the bomb bursts seen under the nose of this USS Boxer based jet. Trim color was Yellow. 4 July 1952. (National Archives)

F9F-4 (125943) of the Checkerboard squadron VMF-312 is seen above in the standard Glossy Sea Blue finish and White lettering. The trim around the tailpipe is Red outlined with White. The 943 appears in Black on the Natural Aluminum tip of the wing tank. LaCrosse, Wisconsin. Fall 1953. (Bob Stuckey)

F9F-4

The F9F-4 first flew in July of 1950 and represented an improved and slightly larger version of the Panther. Its gross weight without external stores topped out at 17,500LBS (a full load of external stores would run that weight up by 3,465 additional pounds). Total fuel capacity was 1003 gallons. The straight fin and rudder and slightly lengthened fuselage are the external details that make for easy identification of this variant. The F9F-4 was powered by the Allison J33-A-16 engine with 6,450LBS of thrust and was armed with four 20MM cannon. Launch stubs for bombs and rockets were provided. Top speed was 575MPH. There were 109 F9F-4s produced, most being assigned to USMC squadrons some of which saw combat in Korea.

F9F-4 (125226) from MAG-33 is seen on the Marston mat steel strip at Pohang, Korea (K-3). Note the extended dive brakes, the natural aluminum strip on the fin and horizontal stabilizer leading edges. Trim color on the nose is Yellow. September 1953. (Clay Jansson)

Lt. Davis and Col. Coss make a pre-flight inspection of their VMF-115 F9F-4s. (125164 & 125226). The squadron departed within minutes for North Korean targets. 15 March 1953. (USMC)

F9F-4 General Arrangement

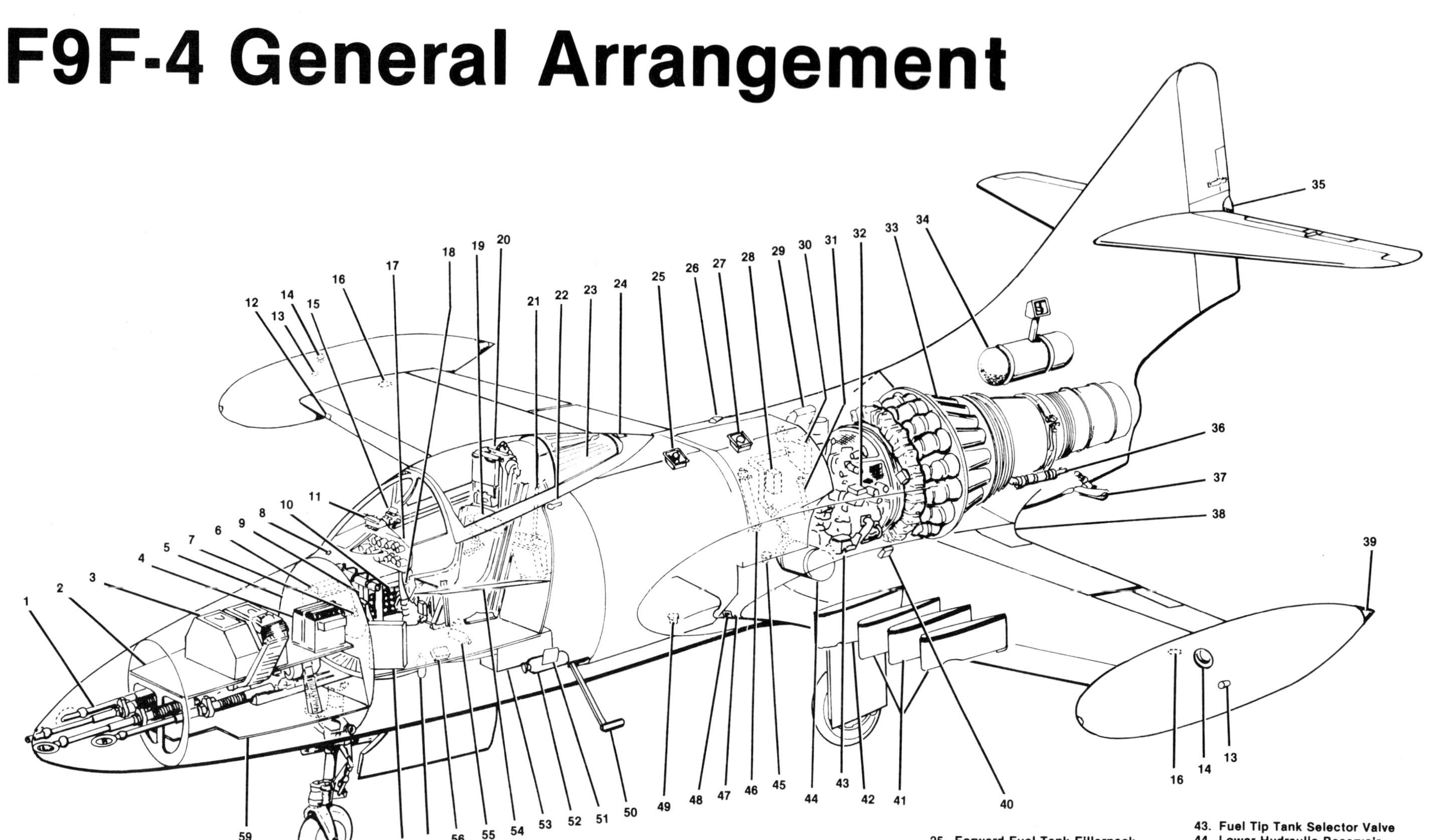

1. 20mm Guns (4)
2. Radio Equipment Deck
3. Inboard Guns Ammunition Boxes
4. Battery
5. Forward Armor Plate
6. Right Cockpit Distribution Box
7. Hydraulic System Accumulator
8. Barrier Guard
9. Rudder and Brake Pedals
10. Circuit Breaker Panel
11. Gun Camera
12. Auxiliary Wing Position Light
13. Wing Tip Position Lights
14. Wing Tip Tank Filler Cap
15. AFCS Gunsight
16. Formation Lights
17. Instrument Panel
18. Control Stick
19. Oxygen Bottle
20. Ejection Seat Face Cover Handles and Head Rest
21. Cabin Pressurizing and Cooling Turbine Oil Fillerneck
22. Canopy Control (for ground operation)
23. AN/ARN-6 Sense Antenna
24. Canopy Unlatch (for removal on ground)
25. Forward Fuel Tank Fillerneck
26. Top Fuselage Light
27. Rear Fuel Tank Fillerneck
28. Generator Connector Box
29. Upper Hydraulic Reservoir
30. Engine Junction Box
31. Engine Oil Fillerneck
32. Engine Accessories
33. J33-A-16 Engine
34. Water Injection Fluid Tank
35. Tail Position Lights
36. Arresting Hook Recoil Strut
37. Tail Skid
38. Bomb Clearance Flap Door
39. Wing Tip Tank Dump Valve (2)
40. D-C External Power Receptacles
41. Rocket Launchers (6)
42. Bomb Racks (2)
43. Fuel Tip Tank Selector Valve
44. Lower Hydraulic Reservoir
45. Fuel Tank Water Drain
46. Aileron Booster Hydraulic Accumulator
47. Wing Lock Indicating Rod
48. Approach Light
49. Fuel System Drain Valve
50. Boarding Ladder
51. Handgrip and Step (ladder release)
52. Landing Gear Emergency Air Bottle
53. Outboard Guns Ammunition Boxes
54. Left Cockpit Console
55. Canopy Emergency Air Bottle
56. A-C External Power Receptacle
57. Brake and Seat Ejection Emergency Air Bottle
58. Left Cockpit Junction Box
59. Gun Deck

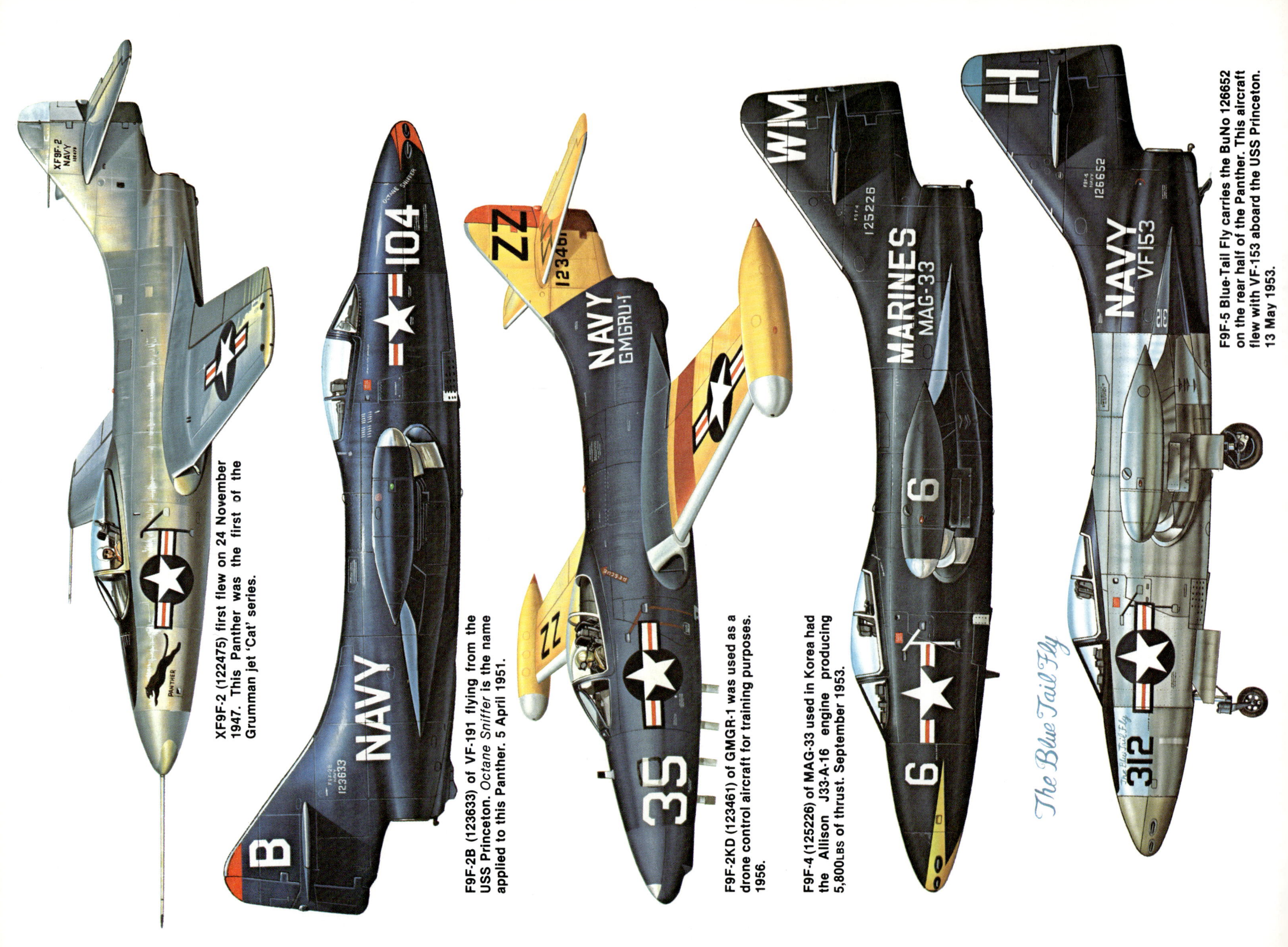

XF9F-2 (122475) first flew on 24 November 1947. This Panther was the first of the Grumman jet 'Cat' series.

F9F-2B (123633) of VF-191 flying from the USS Princeton. Octane Sniffer is the name applied to this Panther. 5 April 1951.

F9F-2KD (123461) of GMGR-1 was used as a drone control aircraft for training purposes. 1956.

F9F-4 (125226) of MAG-33 used in Korea had the Allison J33-A-16 engine producing 5,800Lbs of thrust. September 1953.

F9F-5 Blue-Tail Fly carries the BuNo 126652 on the rear half of the Panther. This aircraft flew with VF-153 aboard the USS Princeton. 13 May 1953.

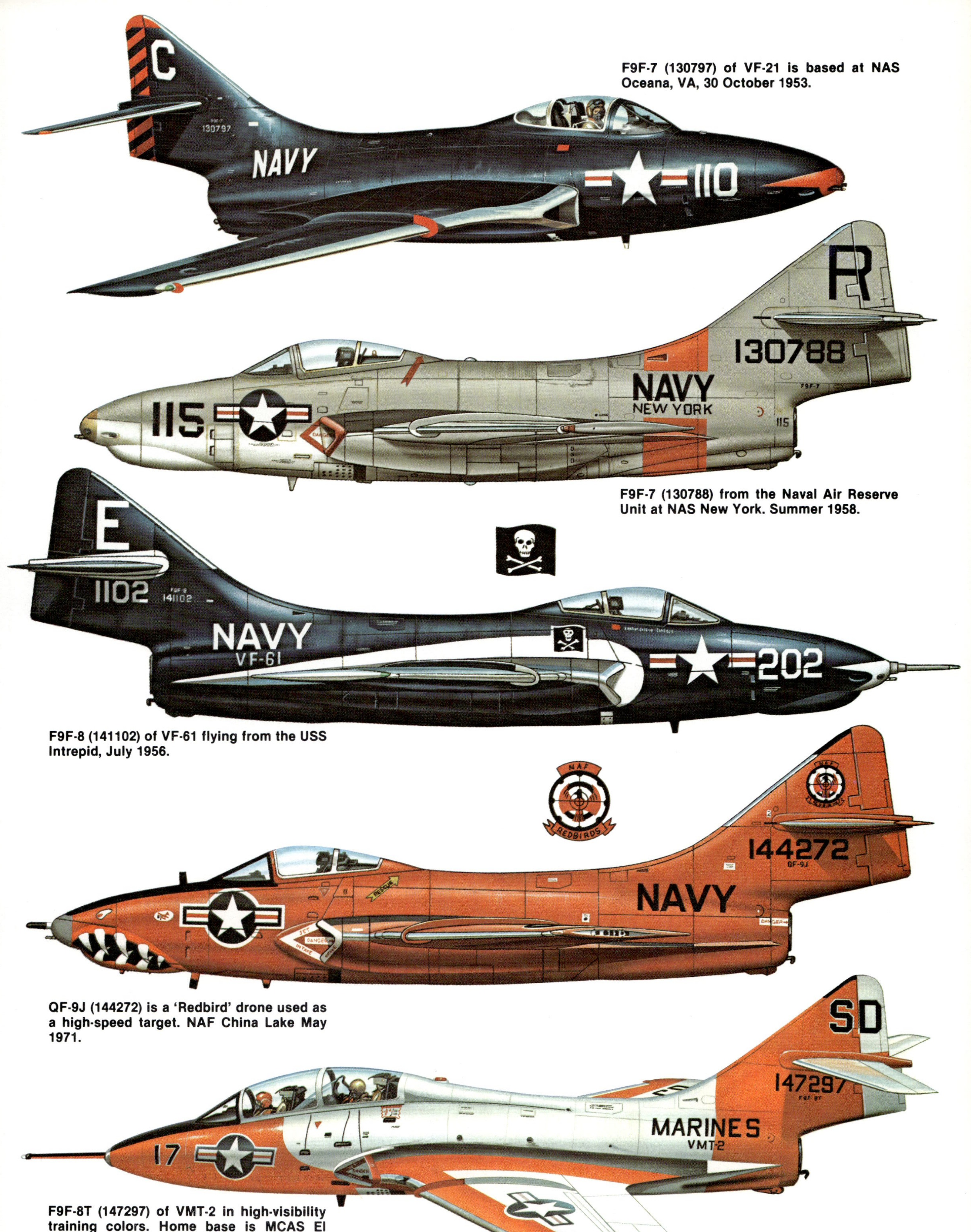

F9F-7 (130797) of VF-21 is based at NAS Oceana, VA, 30 October 1953.

F9F-7 (130788) from the Naval Air Reserve Unit at NAS New York. Summer 1958.

F9F-8 (141102) of VF-61 flying from the USS Intrepid, July 1956.

QF-9J (144272) is a 'Redbird' drone used as a high-speed target. NAF China Lake May 1971.

F9F-8T (147297) of VMT-2 in high-visibility training colors. Home base is MCAS El Toro, CA, 15 September 1959.

Miami based F9F-4s of VMA-334 fly off the South Atlantic coast. All aircraft show the weathering effects of the blistering hot Florida sun. All lettering is White, and the nose, tip tanks and fin tip carries Red trim with a White outline. Late 1953. (Clay Jansson)

(Above) F9F-5 (126251) of VMA-224 flies over the countryside near MCAS El Toro, CA. The trim colors on the nose and tail are Red outlined in White. Each of the two wing tip tanks held 120 gallons. 1956 (Clay Jansson via Peter Mancus)

F9F-5

The F9F-5 was externally identical to the F9F-4 and was easily recognized from other earlier versions by the taller and straighter tail. The internal difference being the installation of a Pratt & Whitney J48-P-6 engine producing 6,280LBS of thrust (approximately 12,000HP) and provided a speed of 579MPH at 5,000FT. The gross weight of a fully fueled F9F-5 was 21,245LBS and Grumman listed the range as 1,130 miles with a service ceiling of 42,800FT. The armament and ordnance arrangements were the same as the F9F-2/3s. To improve the handling at lower speeds and during landings, wing fences were installed. The first flight was on 21 December 1949 and during the production run, Grumman produced 616 F9F-5s. A small number of these and a few F9F-2s, were delivered to the fleet without their standard Glossy Sea Blue paint...results of this testing indicated painting to be a more satisfactory manner of corrosion control.

A pair of 500LB high explosive bombs will be loaded on F9F-5 (126169) of VMF-115 at K-3 Korea. September 1953. Note the name *Patsy* on the nose. (Clay Jansson)

Both catapults having been fired, F9F-5 (125550) of VF-102 is guided on the starboard launch track aboard the USS Tarawa. The Panther just airborne is departing at approximately 110MPH. March 1954. (Walt Olrich)

F9F-5 (125572) of VF-113 aboard the USS Boxer off San Diego, CA. The wing folding mechanism was designed with strength capable to operate the wing with full tip tanks and six 500LB bombs in a 32 knot wind across the deck. Nose and tail trim on this Panther is Medium Blue. 25 February 1953. (National Archives)

F9F-5 of MARS-37 has collapsed the starboard landing gear and dug up some Florida real estate after leaving the runway at NAS Miami, Fla. October 1953. (Clay Jansson)

Seen during a two week deployment, F9F-5s of VMA-223 are parked on the flight line at NAS Whidbey Island. The sharkmouth Panther in the foreground is BuNo 126013. August 1955. (National Archives)

The Blue Tail Fly was a unique combination of two different F9F-5s. Both aircraft belonged to VF-153 aboard the USS Princeton. The story of the Blue Tail Fly goes that a Lt. Richard (Stretch) Clinite was flying an experimental-finish F9F-5 when he was hit by Communist flak and the tail section of the aircraft was badly riddled but he managed to get the Panther back aboard. In the meantime, Ens. William Wilds, Jr. brought back a heavily damaged standard Glossy Sea Blue finish F9F-5 with its rear section intact. After mating the two airframes, the Blue Tail Fly was born. It flew 12 missions in that configuration before being returned to the States for rebuilding.

Wing Fence

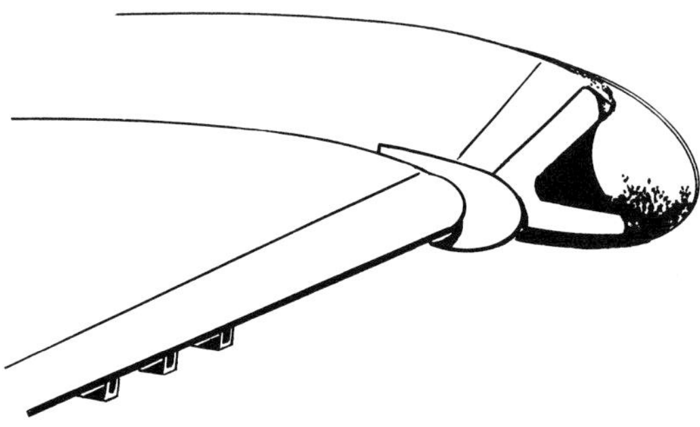

(Left) F9F-5 (125644) of VF-111 from the USS Boxer heads for the Korean coast. Natural aluminum finish with Black lettering. 14 June 1953 (National Archives)

(Below) F9F-5 (126089) of VF-82 lands aboard the USS Antietam off the coast of Cuba. The pilot is receiving the signal to fold wings. 9 February 1953 (National Archives)

(Below) F9F-5 (125449) of VF-94 taxis out toward the Starboard catapult. Carrying a bomb and a rocket under each of the unfolding wings, this Panther is aboard the USS Hornet off Formosa. Trim color on nose and tail is Yellow. 8 September 1954. (National Archives)

Cdr. George C. Duncan Survives Crash

Cdr. Duncan flying F9F-5 (125228) was performing carrier suitability tests for the Patuxent River Naval Air Test Center on 23 July 1951. While landing aboard the USS Midway and just moments from touchdown, his Panther jet sunk low and slammed into the ramp at the end of the flight deck impacting the aircraft just to the rear of the cockpit under the wing leading edge. Instantly the Panther was surrounded by a flaming inferno. The forward portion of the fuselage rolled down the deck with Cdr. Duncan strapped inside. This spectacular series of Navy photographs shows the rescue and firefighting teams rushing to the smoldering hulk hoping for the slim chance that Duncan would still be alive. Their efforts paid off as the thoroughly battered pilot was pulled out of the wreckage, still alive. Remarkably Cdr. Duncan survived and recovered. The aft section of the crashed Panther ended up burning fiercely on the stern gun platform under the flight deck. It was August before a replacement F9F-5 completed the carrier suitability tests. (National Archives via Art Schoeni)

Ens. A.M. Larsen straps into a F9F-5 of VF-721 (a reserve squadron from NAS Glenview, ILL.) aboard the USS Kearsarge off Korea. Larsen's Panther is carrying a full load of rockets. 30 September 1952. (National Archives)

F9F-5 (126043) carried out tests at the Naval Ordnance Test Station at China Lake, CA. The name *Rudolph* decorates the nose of this Panther. The multi-rocket pods on the wing racks are designed to hold four 5IN rockets each. The arrangement here allows the F9F-5 to carry 16 of these rockets vs the standard load of six. The remaining two launch stubs were available to carry additional ordnance. Only slight modification was necessary to achieve this much-improved firepower. November 1955. (National Archives)

A view of the cockpit as seen from the left side of an F9F-5 (126020). This Marine Corps Panther was based at MCAS El Toro, CA. 29 March 1955. (National Archives)

Rudder Development

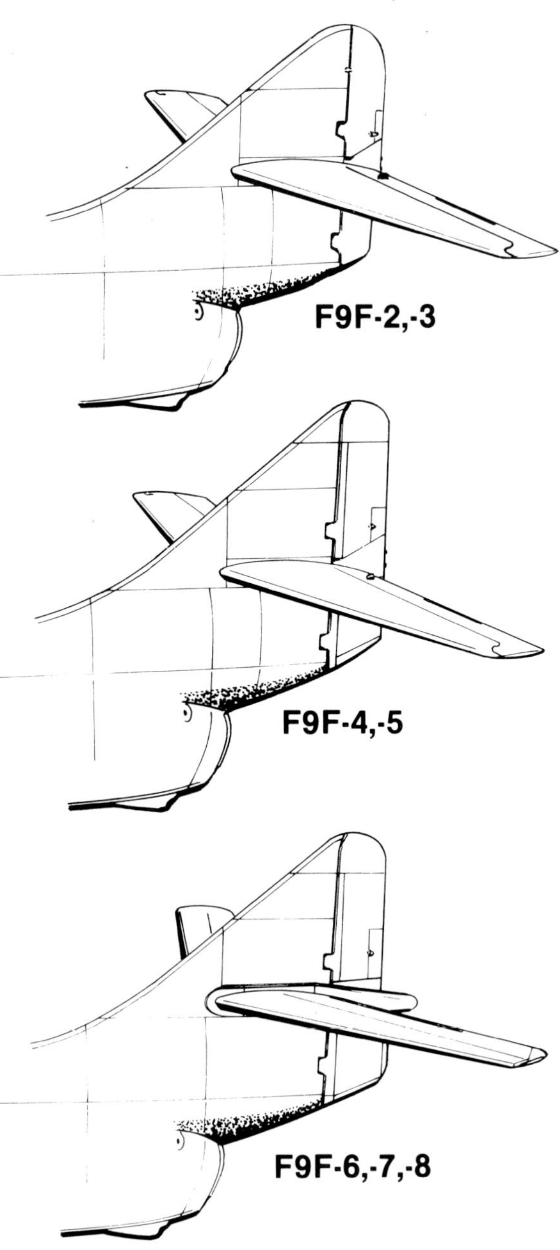

F9F-5KD (126265) tied down on the ramp at the Naval Pacific Missile Range at Pt. Mugu, CA. 1958. (Peter Bowers)

F9F-5KD

A small number of F9F-5Ps (the photo version) were modified by the Navy to a Drone-Controller status by removing the camera equipment and installing radio and radio control gear. Many of these aircraft were used in support of the Regulus missile development program. Eventually as the Panther was replaced by Fleet aircraft of superior performance, some of the last remaining F9F-5KDs were expended as targets themselves. After 1962 these aircraft were redesignated DF-9Es.

F9F-5KD (126277) of VU-1 is recently retired and is seen in the ownership of Ed Maloney at the Ontario Air Museum in 1965. (Clay Jansson via Peter Mancus)

F9F-5P

Grumman produced a total of 36 of the photo-Panthers, the F9F-5P. They were powered by a J48-P-6A engine that produced 6,250LBS of thrust. The photo version had a General Electric G-3 autopilot installation that was required to be engaged for reasons of stability and control during photo operations. In addition to the modification which permitted the attachment of a pair of 150 gallon external tanks on the inboard wing racks, the fuselage forward of the cockpit was lengthened 12 inches for installation of groups of cameras for general reconnaissance, reconnaissance mapping and beach reconnaissance. A photographic viewfinder was installed just forward of the cockpit. Camera positioning and operation of the viewfinder were completely automatic and controlled from the cockpit. The performance of the F9F-5P was virtually identical to that of the F9F-5.

F9F-5P (126279) sparkles in its Glossy Sea Blue factory finish. The wing racks hold external tanks with a total of 300 extra gallons of fuel. The other six wing racks have been removed (as they were on all F9F-5Ps). 1950. (Grumman)

F9F-5P
Camera Installation

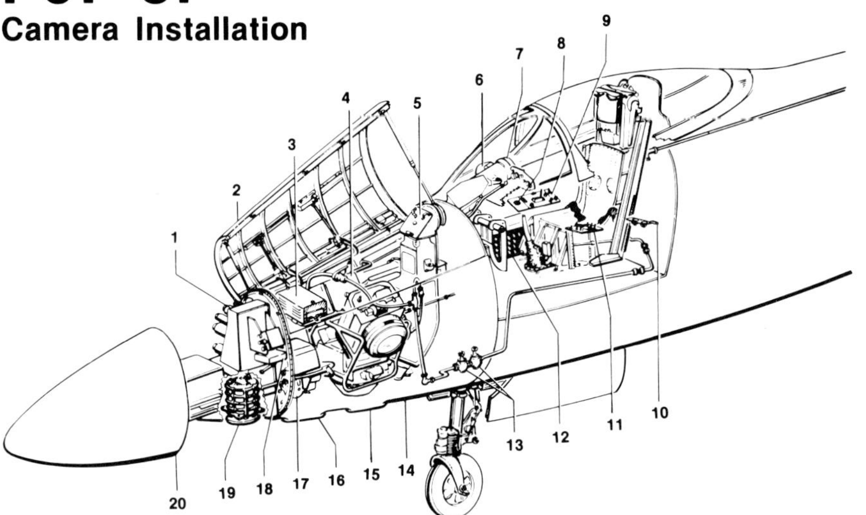

1. Photo Recorder
2. Camera Access Hatch
3. Sonne Amplifier
4. Trimetrogon Camera in Aft Bay
5. Viewfinder
6. Viewfinder Controls and Indicators
7. Viewfinder Eyepiece
8. Interval Computer (Multi-Camera Interval Control)
9. Right Console - Auto Pilot Controls
10. Camera Compartment Heat Control
11. Left Console - Camera Controls
12. Circuit Breaker Panel
13. Relief Valves
14. Viewer Compartment
15. Aft Camera Bay Window
16. Forward Camera Bay Window
17. K-17-6 in. Camera in Fwd Bay
18. Photo Recorder Gun Camera
19. Scanner
20. Sliding Nose Section

F9F-5P (126277) is directed to the starboard catapult during VMCJ-3 carrier qualifications aboard the USS Hornet off the coast of Southern California. October 1956. (Clay Jansson via Peter Mancus)

F9F-5P (126281) of VMCJ-3 breaks out of some low clouds over the mountains near MCAS El Toro, CA. The color trim on the nose, rudder and wing tip tanks is Red and White. c.1956. (Clay Jansson)

(Right) F9F-5P (126287) of VMJ-3 is parked and tied down during a stopover at NAS Anacostia. All markings are White including the twin rings on the nose and wing tip tanks. 16 April 1953. (National Archives)

(Below) Based at MCAS Miami, Fla., a VMJ-3 F9F-5P (126277) wings its way over the Florida coast line. May 1954. (Clay Jansson via Peter Mancus)

F9F-6 Cougar

While maintaining the sleek Panther fuselage lines, the Grumman engineers integrated a sleek 35 degree swept wing and 'flying' tail and came out with a new plane, the 'Cougar'. The F9F-6 was the first of that series with 646 of them being built. The power plant was a Pratt & Whitney J48-P-8 engine delivering 7,250LBS of thrust. Including the 78 gallon tank in each wing, 1,202 gallons of fuel could be carried. The fuselage speed brakes were maintained on this and all other Cougar versions. The armament continued to be four 20MM cannon. The prototype Cougar (126670) first flew in September of 1951. Although aboard at least one aircraft carrier operating off Korea, Cougars saw no combat as the signing of the Peace Treaty on 27 July 1953 halted military operations.

(Above) F9F-6 (127367) of VF-24 taxis after landing aboard the USS Yorktown. Nose and rudder trim is Red. April 1953. (National Archives)

F9F-6 (130929) of VF-191 is trimmed in Red and White. Note the bulge under the nose that housed the IFF radar. San Francisco, CA. 8 August 1954. (Bill Larkins)

Wing Development

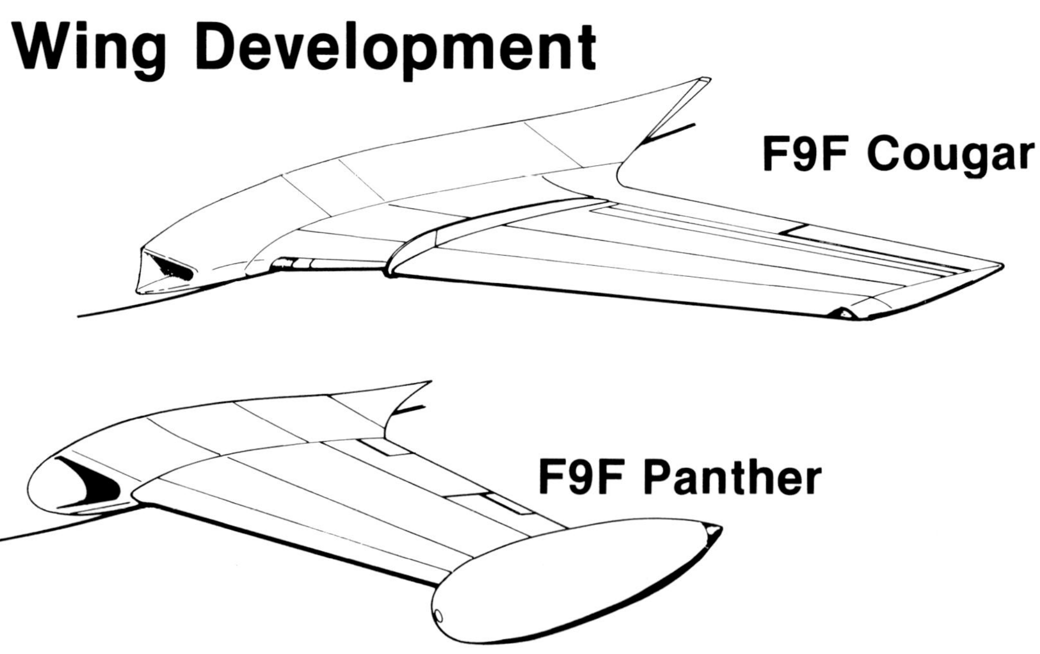

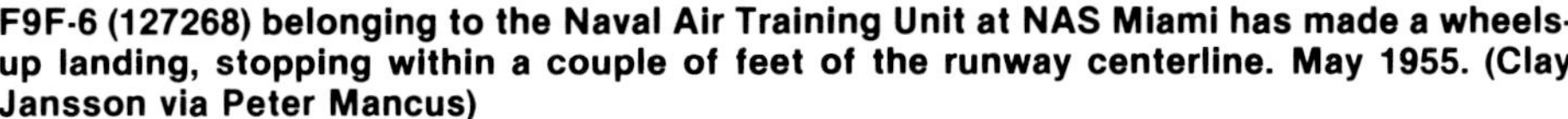

F9F-6 (127268) belonging to the Naval Air Training Unit at NAS Miami has made a wheels-up landing, stopping within a couple of feet of the runway centerline. May 1955. (Clay Jansson via Peter Mancus)

(Top Right) Structural damage was received by this F9F-6 (127339) of VF-142. During carrier qualifications aboard the USS Hancock, the Cougar went through the wires and into the barrier. 30 March 1955. (National Archives)

(Right) F9F-6 (127419) was assigned to VMFT-10 at MCAS El Toro. The nose trim was Green with a White outline. 1957. (M/Sgt. W.F. Geminhardt)

(Lower Right) With the 7th Fleet off Formosa, F9F-6 (127340) was assigned to VF-91 aboard the USS Hornet. Trim was Red outlined with White. 5 September 1954. (National Archives)

Lt. B.L. Havron brings his F9F-6 of VF-122 aboard the USS Shangri La. 1957. (National Archives)

(Left) F9F-6 (128144) of VF-33 carries a Yellow lightning bolt marking just over the White nose number. These Cougars were assigned to the 6th Fleet aboard the USS Midway. February 1954. (National Archives)

(Below Left) F9F-6D (128246) is a surviving drone aircraft from Naval Ordnance Test Station at China Lake, CA. seen at the Naval storage facility at Litchfield Park. 21 March 1960. (William Swisher)

(Below) Many Cougars saw service with reserve units. F9F-6 (127266) from NRAB Lincoln, NE is seen at St. Louis, MO in June 1957. (Bill Larkins)

F9F-6D/-6K2

The Cougar was the first swept-wing high-speed drone. It was modified by the Navy from the standard F9F-6 airframe with the installation of specialized radio equipment. These changes resulted in the F9F-6D which was used as a drone director and the F9F-6K2 which could be used as a drone-controller or target itself. Externally only the antenna arrangement and colorful Red, Yellow and Blue paint scheme differed from the 'standard' F9F-6.

F9F-6P (128256) of VMJ-2 demonstrates fully extended dive brakes during flight from its home field at MCAS Cherry Point, NC. 1955. (M/Sgt. W.F. Gemeinhardt)

F9F-6P

A total of 60 F9F-6Ps were turned out by Grumman as photo reconnaissance aircraft. Using the same airframe and powerplant as the F9F-6, all four guns were removed and replaced with nose-mounted trimetrogon, K-17 (6 inch) and photo recorder gun cameras. Entering Fleet service in early 1953, photo-Cougars served until the late '50s. A number of these received the new Grey/White paint scheme when the Navy changed from the Glossy Sea Blue in 1957.

F9F-6P
Camera Apertures

(Above) F9F-6P (128310) of VMJ-2 has White chevron trim on its Glossy Sea Blue finish. MCAS Cherry Point, NC. 1956. (M/Sgt. W.F. Gemeinhardt)

(Center Left) F9F-6P (127488) first served with VC-61 before being assigned to the Navy Technical Training Unit at NAS Pensacola, FL. This Cougar was stricken from service in December 1959. Photo taken at Minneapolis/St. Paul, MN. 1958. (Bob Stuckey)

(Left) Passing off the bow of the USS Randolph, F9F-6P (128295) of VC-61 prepares to break and enter the landing pattern. 29 April 1954. (National Archives)

F9F-7

Externally identical to the F9F-6, a total of 168 F9F-7s were produced by Grumman. As with the F9F-3 and F9F-4, the F9F-7 was Allison powered. A 6,350LB thrust J33-A-16 engine delivered a top speed of just over 650MPH. The cockpit layout though basically similar, was instrumented to operate only as a day fighter. The leading edge flaps were tied in hydraulically with the flaps so they would act in unison for better handling characteristics particularly during landing, much the same as the earlier versions.

(Above) F9F-7 (130872) assigned to the Naval Air Reserve Training Unit at NAS Olathe, Kansas. In February of 1957 it was sent to NAS Norfolk where it was converted to a drone (F9F-6K2). It survived until June of 1961. NRAB Minneapolis, MN. Winter 1955. (Bob Stuckey)

(Below) F9F-7 (130831) assigned to the Dallas reserves is seen during stopover at Wilmington, NC. Note the international orange fuselage band and the position of the steerable nosewheel. 27 July 1955. (Paul J. McDaniel)

One of two Cougars modified to take part in an experiment called 'Flex Deck'. F9F-7 (130863) has landed on a rubberized fabric 'deck' which was secured to air bags. The 'deck' was 375FT long with arresting cables stretched across its 80FT width. The purpose of this Navy experiment was to determine the practicality of saving weight by eliminating the need for landing gear. NATC Patuxent River, MD. February 1955. (National Archives)

A flight of four sleek F9F-7s in the colorful Red markings of VF-21 fly over the Atlantic out of their home base at NAS Oceana, VA. Cougar 105 is BuNo 130802. 30 October 1953. (National Archives)

F9F-8 (138830) of VA-126 carries a single multiple-launch rack beneath each wing. The rudder trim colors are Dark Green and White. NAS Miramar, CA. 10 August 1957. (Bill Larkins)

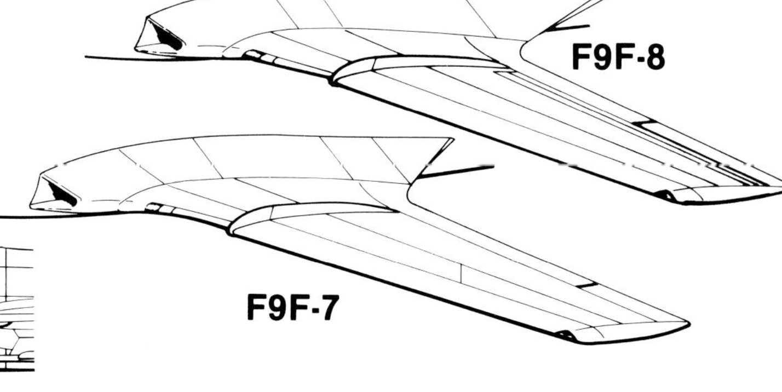

Armed to the teeth, this F9F-8 (141173) of VX-3 carries four sidewinder missiles. March 1956. (National Archives)

F9F-8

The F9F-8 series represented the last of the Cougar line. Grumman produced 601 of the F9F-8s. These aircraft featured a change in the wing design making it larger and providing for six underwing racks for ordnance or external fuel tanks. Provisions were made for an in-flight refueling probe located on the tip of the nose section, an extended tail and increased fuel capacity. The first production model flew on 18 January 1954. The powerplant installed was a Pratt & Whitney J48-P-8A engine producing 7,250LBS of thrust with a top speed of 714MPH and a range of 1,000 miles. The four 20MM cannon armament was retained. Although sub-sonic, the F9F-8 could safely exceed Mach 1 in dives. Based on its performance, the Blue Angels adopted it in 1955.

Nose Development

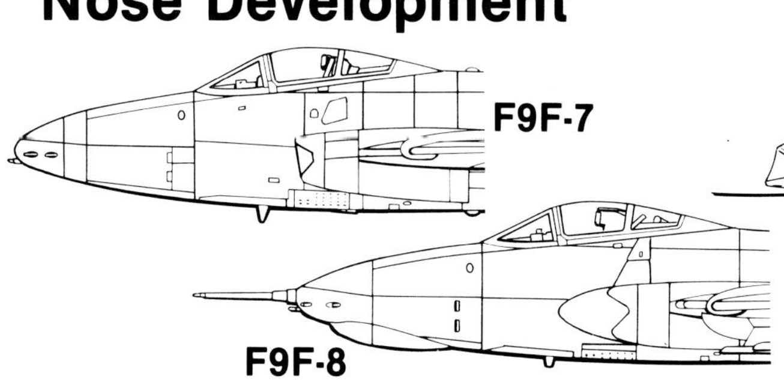

F9F-8 (141655) of FASRON-4 in Grey/White finish with Medium Blue trim on the tail. Moffett Field, CA. 17 May 1958. (Bill Larkins)

Wing Development

F9F-8 (141065) of VF-112 looses its nose section forward of the canopy on impact with the water. USS Essex off San Diego, CA. May 1956. (National Archives)

F9F-8 Cougar
Nose Section Development

F9F-8,-8B

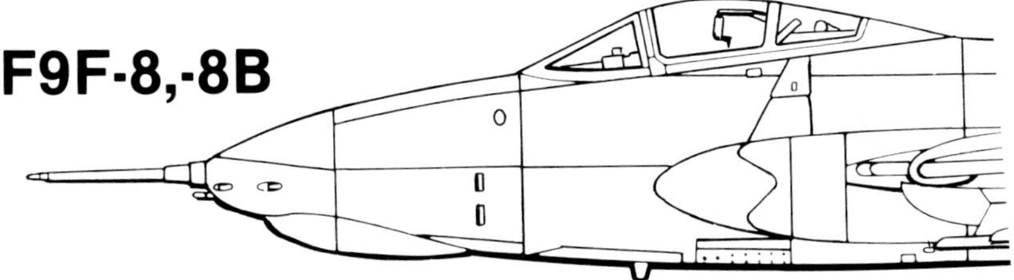

F9F-8P

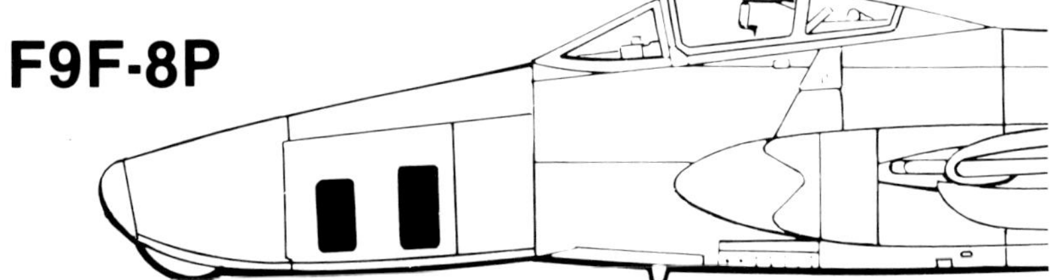

F9F-8T

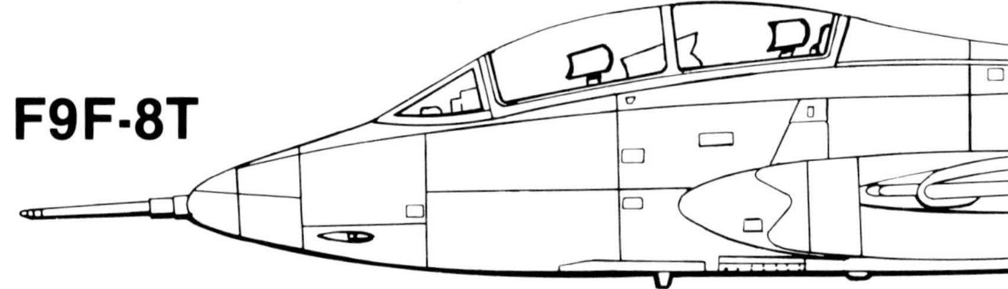

F9F-8KD

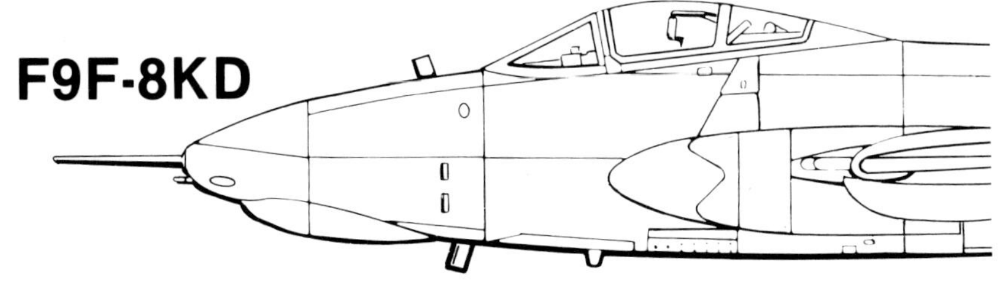

A flight of four F9F-8s of VF-123 demonstrate multiple in-flight refueling while attached to a R3Y-1 flying boat. 31 August 1956. (National Archives)

F9F-8 (131251) of VF-93 sits on the ramp at NAS Oakland, CA. Fuselage trim is Medium Blue with a White band. The name under the cockpit is LT JG CAUSEY. 18 September 1955. (Bill Larkins)

F9F-8B (131069) of FASRON-8 is piloted by Lt. Walt Ohlrich near NAS Alameda, CA. 28 September 1958. (Walt Ohlrich)

Trimmed in Red, F9F-8B (141057) of VMA-533 is parked on the flight line at MCAS Cherry Point, NC. 1958. (M/Sgt. W.F. Gemeinhardt)

F9F-8B

Modified from F9F-8s with the Low Altitude Bombing System, internal changes made the difference in the F9F-8B. Changes involved installation of dive and roll indicator on instrument panel, zeroing of dive and roll indicator with ACS-AERO 18A rack outside a/c, installing accelerometer, horizontal yaw-roll gyro, vertical gyro and the installation of a relay box rack in the nosewheel well. These modifications were made by the Navy.

F9F-8B (141156) of VMFT-10, a training squadron based at MCAS El Toro, CA. 1958. (M/Sgt. W.F. Gemeinhardt)

F9F-8B (138855) of VMA-121 in flight over its North Carolina base at MCAS Cherry Point. This aircraft is equipped with LABS for loft bombing with Atomic stores up to 2,000LBS. After 1962 these F9F-8Bs were redesignated AF-9Js. 1958. (M/Sgt. W.F. Gemeinhardt)

F9F-8B (131234) proudly wears the faded Sea Blue finish and White markings of VMF-114. Tailhook is alternating bands of Black and White. Based at Cherry Point, this Cougar was seen in Wilmington, NC. 22 May 1957. (Paul J. McDaniel)

F9F-8P

The final photo variant, the F9F-8P was powered by a Pratt & Whitney J48-P-8A engine developing 7,250LBS of thrust. With the cannon removed and all cameras installed, the performance of the F9F-8P was almost identical to that of the F9F-8. The nose of the Cougar forward of the cockpit was elongated to allow room for the installation of no less than 14 cameras with the capability of photography from the front, the left and right and directly beneath. The cockpit had a viewfinder in the location normally used for the gunsight. 110 of the photo version F9F-8P were built by Grumman.

F9F-8P (144382) of VMCJ-3 banks out over the Pacific ocean. This photo-Cougar is home-based at MCAS El Toro, CA. August 1957. (Clay Jansson via M/Sgt. W.F. Gemeinhardt)

The ground crew makes final preparation for a flight. This F9F-8P belongs to VC-62. April 1956. (National Archives)

General Arrangement

1. CA-13B Camera with CIL-8 Cone
2. CA-17A Camera Integral Lens Cone
3. Dehydrator and Motor
4. Scanner Converter, Mod HA
5. System Relay Unit
6. Forward Station Relay Unit
7. Overrun Control
8. Window Washing Tank and Filler
9. VF-34 Viewfinder
10. Station Relay Units, AB
11. Servo Power Unit
12. Central Computer
13. Scanner
14. Battery
15. Magazine Relay Unit
16. Rotary Mount STA 3
17. Camera Window STA 3
18. Rotary Mount STA 2
19. Camera Window STA 2
20. Camera Mount STA 1
21. Access Door STA 1
22. Forward Camera Window

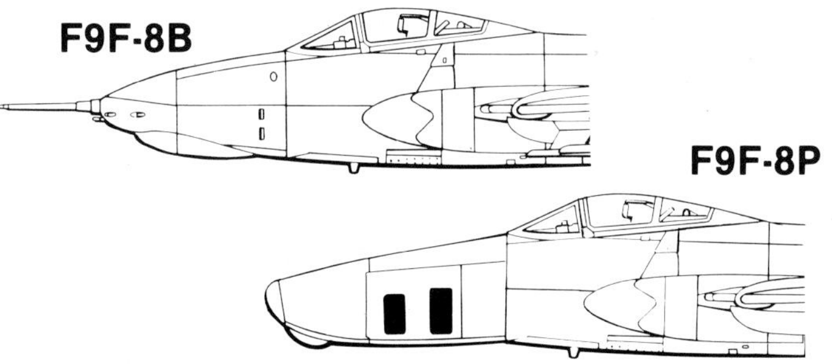

F9F-8P (141710) of VFP-61 in the standard Grey/White scheme with Red trim. This colorful photo-Cougar was photographed at NAS Miramar, CA. 10 August 1957. (Clay Jansson via Peter Mancus)

Nose Development

F9F-8B

F9F-8P

(Center) In-flight refueling is in progress between F9F-8P (144383) of VMCJ-3 and an AJ-2 Savage tanker (134042) of VAH-16. This action over the Sea of Japan during the Summer of 1959. (Clay Jansson via Peter Mancus)

F9F-8P (141726) of VFP-61 attached to Carrier Air Group 14. This Cougar seen at NAS Miramar, CA. Markings are Orange-Red. Note Fox painted on landing gear door. 10 August 1957. (Bill Larkins)

TF-9J (147362) of RVAH-3 in the trainer White scheme with Orange-Red trim. Seen at the Naval Air Facility at Andrews AFB, MD. May 1965. (Clay Jansson)

F9F-8T

Redesignated in 1962 to TF-9J, the two-seat version was the last of the line of Grumman Cougars. The fuselage was lengthened 34 inches to accomodate the second seat. To help compensate for the additional weight, two of the four 20MM cannon were removed. The power was provided by the standard engine for this series, the J48-P-8A. Provisions were made for carrying two 150 gallon external fuel tanks. The first F9F-8T flight was on 29 February 1956. Four of these Cougars made it to Vietnam and participated in the action there working as Forward Air Control a/c for the Marines with H&MS-11 flying out of Da Nang in late 1966.

(Top Right) F9F-8T (142470) low over the water of Pensacola Bay during a show. Even after the Blue Angels traded in their Cougars, a F9F-8T was kept to use for press flight demonstrations. April 1958. (USN via Tom Curry)

(Center Right) F9F-8T of HAMRON-15 flies over the rugged California terrain near its MCAS El Toro base. Cougar is White with Orange-Red and Black markings. 1958. (Clay Jansson)

(Bottom) One of the four TF-9Js assigned as a FAC a/c by H&MS-11 in Vietnam is seen at NAS North Island, CA. 15 November 1967. (Clay Jansson)

F9F-8T (142448) from the Naval Parachute Facility at NAS El Centro, CA. This aircraft had the rear portion of the canopy removed for ejection seat testing. A similar F9F-8T was used at the Naval Air Test Center on 28 August 1957 when a test was done with ground level ejection with a 'live' person punching out. At 125KNOTS with the wheels still on the runway during take off, the demonstration was flawless and the result was better protection for the pilot during take offs and landings. January 1961. (Clay Jansson)

QF-9J

Eventually even the latest of the Cougars, the F9F-8s were replaced by more powerful Naval aircraft. Radio control equipment was added to these Cougars which converted them to high performance drones capable of high-G turns making them more realistic targets.

(Right) QF-9J (144288) in Dayglo Orange finish. This target drone seen at Pt. Mugu, CA. 17 October 1970. (Flight Leader Russell-Smith)

USN/USMC TERRIBLE TWINS OF WWII

BY JIM SULLIVAN WITH **MORE** TO COME . . . .